Instructor's Guide for

An Introduction to Physical Science

Eighth Edition

James T. Shipman
Ohio University

Jerry D. Wilson
Lander College

Aaron W. Todd
Middle Tennessee State University

Houghton Mifflin Company Boston New York

Editor-in-Chief: Kathi Prancan
Assistant Editor: Marianne Stepanian
Editorial Assistant: Katie Shutzer
Senior Manufacturing Coordinator: Lisa Merrill
Executive Marketing Manager: Karen Natale

Printed in the U.S.A.

ISBN: 0-669-41716-5

123456789-VG-00 99 98 97 96

Contents

Preface

The purpose of this *Instructor's Guide* is to facilitate the teaching of physical science from the textbook *An Introduction to Physical Science*, Eighth Edition, by Shipman, Wilson, and Todd. By supplying answers to all end-of-chapter review questions, thought questions, exercises, and relevance questions, and to the multiple-choice and short-answer quizzes, we hope to help both experienced and inexperienced instructors (especially the latter) who are presenting such a multidisciplined course as physical science. Not many instructors so involved are expert in all five sciences—physics, chemistry, astronomy, geology, and meteorology.

How should the course in physical science be taught? Most instructors already have a personal philosophy of teaching that can be applied to teaching physical science. However, we would like to mention a few of the tenets that have determined how physical science has been taught in our classes.

We have been most successful in our teaching when the concepts presented are relevant to the student's environment and experience. We have done this in the textbook whenever possible. We emphasize to our students the importance of mathematics and symbol notation. Many of the students in our classes will become elementary school teachers, and we believe that they should be aware of the role of mathematics in the sciences. In fact, we think it important that all physical science students be able to participate with some degree of success in the quantitative aspect of the sciences. However, the amount of math presented is at the discretion of the instructor, and the mathematical sections need not be stressed.

The interests and mathematical skills of physical science students vary over a wide range. Since the study of physical science is best described by numbers, a certain amount of math is necessary in presenting our concepts of the physical world. Thus, instructors adjust the level of math to the abilities of the students. Appendixes II through V in the textbook explain the basic mathematical skills, and contain practice problems with answers so that the students can test themselves.

Instructors have varied opinions concerning the treatment of math and its importance in the teaching of science. Some believe strongly in the use of math. Others believe in using math only when necessary. We have treated the concepts both descriptively and quantitatively, with the relative emphasis on each of these two approaches left to the discretion of the instructor.

SAMPLE TEACHING PLANS

A one-year physical science course can be taught on either the semester or the quarter plan. Possible breakdowns of how a one-year, three-hour course can be taught are given below. In the semester plan, physics and chemistry are taught the first semester, and astronomy, meteorology, and geology the second semester. This order can be switched, if desired—that is, Earth sciences can be taught the first semester, and the physics and chemistry the second semester. Under the quarter plan, classical physics and astronomy are taught the first quarter, modern physics, and chemistry the second quarter, and meteorology and geology the third quarter. This ordering can be switched; for instance, the second and third quarters can be reversed.

Semester Teaching Plan

First Semester		Second Semester	
Chapter	Number of Lectures	Chapter	Number of Lectures
1	3	16	5
2	3	17	4
3	3	18	4
4	3	19	4
Exam		Exam	
5	3	20	3
6	3	21	3
7	3	22	3
Exam		23	3
8	3	24	4
9	3	Exam	
10	3	25	4
Exam		26	4
11	3	Exam	
12	3	(final)	
13	3		
14	3		
15	3		
Exam			
(final)			

ORGANIZATION OF THIS INSTRUCTOR'S GUIDE

This guide has been written on a chapter-by-chapter basis, corresponding to the chapters in *An Introduction to Physical Science*, Eighth Edition. Most chapters contain all of the following:

Introduction
Demonstrations
Answers to Review Questions
Answers to Critical Thinking Questions
Answers to Exercises
Answers to Relevance Questions
Answers to Study Guide Quiz (multiple-choice questions; short-answer questions)

The introduction presents the concepts that we believe can be emphasized, expanded upon, or omitted, in addition to general information. The sections containing answers to the questions and exercises from the text will be particularly useful to the nonspecialist. All questions in the textbook are answered and the exercises are solved step by step. Instructors who assign the quizzes at the end of each *Study Guide* chapter will find the answers included here. Finally, a variety of teaching aids and references for further reading are listed at the back of this *Guide*.

Other elements in the Eighth Edition package are the *Laboratory Guide*, the *Instructor's Resource Manual to the Laboratory Guide*, the *Study Guide* for students, and transparencies of artwork in the text. The *Instructor's Test Bank* is available for IBM, Apple, and Macintosh, as well as in a print version. It contains multiple-choice questions, completion questions, and exercises, all arranged chapter by chapter. We hope this *Instructor's Guide*—indeed, the entire instructional package—will help you present physical science to students in a more interesting and stimulating manner. If you have problems, questions, or comments concerning any aspect of the teaching of physical science as presented, please feel free to contact us, either directly or through Houghton Mifflin.

J.T.S.
J.D.W.
A.W.T

Introduction and Chapter 1

Measurement

The introduction reviews the objects and approach to science and gives the student an awareness of his or her place in today's physical environment. The five major divisions of physical science are defined, and an overview of the scientific method is given. It is recommended that the Introduction be assigned reading.

Chapter 1 is very important because all quantitative knowledge about our physical environment is based on measurement. Also, major terms such as *hypothesis*, *concept*, *theory*, and *scientific method* are considered. The idea that physical science deals with quantitative knowledge should be stressed. It is not enough to know that a car is going "fast"; it is necessary to know how fast. Section 1.8, "Approach to Problem Solving," is designed to provide the student better insight and technique in handling quantitative exercises.

The concepts of mass and weight are discussed briefly in this chapter; a more complete discussion is found in Chapter 3. The concept of electric charge is also introduced as a fundamental quantity; it will be discussed more fully in Chapter 8. A good understanding of units is of the utmost importance, particularly with the metric conversion occurring today. The metric SI is introduced and explained. Both the metric and British systems are used in the book in the early chapters for familiarity. The instructor may decide to do examples primarily in the metric system, but the student should get some practice in converting between the systems. This provides a knowledge of the comparative size of similar units in the different systems and makes the student feel comfortable using what may be unfamiliar metric units.

The general theme of the chapter and the textbook is the student's position in his or her physical world. Show the students that they know about their environment and themselves through measurements. Measurements are involved in the answers to such questions as, How old are you? How much do you weigh? How tall are you? What is the normal body temperature? How much money do you have? How much water is in the ocean? How many stars are in the sky? These and countless other questions are resolved or answered by measurements and quantitative analyses. To give students an idea of order of magnitude, the authors usually show the movie *Powers of Ten* the first day of class.

DEMONSTRATIONS

Have a meter stick, a yard stick, a timer, one or more kilogram masses, a one-liter beaker or a liter soda container, a one-quart container, and a balance and scales available on the instructor's desk. Demonstrate the comparative units. The meter stick can be compared to the yardstick to show the difference between them. The liter and quart can also be compared. Pass the kilogram mass around the classroom so that students can get some idea of the amount of mass in one kilogram. Mass and weight may be compared on the balance and scales.

When discussing Section 1.6, have class members guess the length of the instructor's desk in metric and British units. Then have several students independently measure the length with the meter stick and yardstick. Compare the measurements in terms of significant figures and units, and discuss the measurements in terms of accuracy and precision. Compare the averages of the measurements and estimates, and discuss error. Convert the average metric measurement to British units, and vice versa, to practice conversion factors and to see how the measurements compare.

Various metric unit demonstrations are available from commercial sources.

(General references to sources of teaching aids such as films and demonstrations are given in the Teaching Aids section at the back of this *Guide*.)

ANSWERS TO REVIEW QUESTIONS

1. c
2. b
3. Yes. Our senses are our contacts to the physical world.
4. The senses are limited. For example, our eyes are sensitive only to a narrow band of wavelengths of light, and our ears are limited to a small range of physical vibrations. The limitations of smell, taste, and touch are more difficult to define. The temperature sense of touch is limited because of injury.
5. (a) No.
   (b) Yes.
   (c) All the same size.
   (d) Good luck.
6. c
7. d
8. Hypotheses and theories describe phenomena.
9. Basically, any theory must be supported by experiment to be valid.
10. b
11. a
12. Length, time, mass, and electric charge.
13. No, there have to be different events in order to sense or measure time.
14. b
15. d
16. (a) Fraction of terrestrial distance, distance traveled by light.
    (b) Mass of one liter of water, metallic artifact.
    (c) Fraction of average day, in terms of frequency of radiation from cesium-133 atom.
17. The kilogram.
18. The standard unit of time is more basic, because the standard unit of length, the meter, is defined as the distance light travels in one *second*.
19. To establish common standards of measurement for international commerce and science.
20. d
21. b
22. A fundamental quantity is one that provides the foundation for other quantities. Length is an example. A derived quantity is a combination of one or more fundamental quantities. Volume and speed are examples.

23. Density = mass/length^3.
24. kg/m^3 for both.
25. (a) Kilogram of iron.
    (b) Same mass.
26. By using a hydrometer to measure its density.
27. d
28. A comparison of an unknown quantity with a standard unit.
29. Measured numbers always have uncertainty and are not exact. Exact numbers may be measured, but there is an accepted, fixed value.
30. Systematic error arises from a particular measurement instrument or technique and is unidirectional. A random error arises from unknown and unpredictable experimental variations.
31. Accuracy indicates the closeness to the true or accepted value. Precision refers to the spread or dispersion of a series of measurements. Either can be obtained from a set of measurements.
32. Random errors affect precision, and systematic errors affect accuracy.
33. No. Results should be rounded off to the proper number of significant figures in either case.
34. b
35. To avoid writing a large number of zeros.
36. 10^2 or a hundred times.
37. 10^3 or a thousand times.
38. See text.

ANSWERS TO CRITICAL THINKING QUESTIONS

1. The metric decimal system lends itself to easy calculations and conversions through multiples of 10.
2. The liter is larger than the quart, and the kilogram is larger than the pound. However, the kilometer is shorter than the mile.
3. Your choice. Would suggest a decimal system with related subunits. You might honor yourself in naming a unit.

ANSWERS TO EXERCISES

1. (a) 60.0 kg (2.2 lb/kg) = 132 lb; 22 lb
   (b) 60.0 kg
2. (a) 150 kg
   (b) 150 kg (2.2 lb/kg) = 330 lb
   (c) Yes, 330 lb (1/6) = 55 lb
3. (a) 11 lb (1 kg/2.2 lb) = 5.0 kg
   (b) 11 lb (1/6) = 1.8 lb
4. (a) 75.0 lb (1 kg/2.2 lb) = 34.1 kg
   (b) 34 kg
5. 2.0 L (1 kg/L) = 2.0 kg = 2000 g

6. $20\ cm \times 20\ cm \times 30\ cm = 12 \times 10^3\ cm^3$ or mL = 12 L
   12 L (1 kg/L) = 12 kg = 12,000 g

7. 0.085 kg = 85 g (1 mL/g) = 85 mL

8. $350\ cm^3 = 350\ mL = 0.350\ L\ (1\ kg/L) =$ 0.350 kg

9. $\rho = m/V = 500\ g/63\ cm^3 =$ $7.9\ g/cm^3$ (the density of iron)

10. $V = m/\rho = 550\ g/(7.9\ g/cm^3) =$ $70\ cm^3$

11. $m = \rho V = (0.92\ g/cm^3)(3.0 \times 3.0 \times 3.0\ cm^3) =$ 25 g

12. $m = \rho V = (1030\ kg/m^3)(1.25\ m^3) =$ 1.29×10^3 kg

13. 6 ft = 72.0 in. (2.54 cm/in.) = 183 cm = 1.83 m

14. 157 cm (1 in./2.54 cm) = 61.8 in. = 5 ft 2 in.

15. 220 lb (1 kg/2.2 lb) = 100 kg
    6 ft 7 in. = 79 in. (2.54 cm/in.) = 201 cm = 2.01 m

16. 142 cm (1 in./2.54 cm) = 55.9 in. = 4.66 ft = 4 ft 8 in.
    45 kg (2.2 lb/kg) = 99 lb

17. 25 mi/h (1.609 km/mi) = 40 km/h

18. 60 mi/h > 90 km/h. 60 mi/h (1.609 km/mi) = 97 km/h

19. (a) 9.99×10^{-3}
    (b) 6.49×10^3
    (c) 1.06×10^{-2}
    (d) 8.76×10^4

20. (a) 1.1×10^2
    (b) 2.1×10^{-3}
    (c) 9.4×10^3
    (d) 3.4×10^{-4}

21. $(3.2\ m \times 1.04\ m)/0.015\ m =$ 2.2×10^2 m

22. 6.75 (3 sf)

23. (a) 73,000
    (b) 0.000325
    (c) 399
    (d) 0.00234

24. (a) 4.26×10^6
    (b) 2.78×10^3
    (c) 1.02×10^{-2}
    (d) 6.28×10^{-5}

25. (a) 255 Ms = 2.55×10^8 s
(b) 607 km = 6.07×10^5 m
(c) 65 μg = 6.5×10^{-5} g
(d) 0.18 mL = 1.8×10^{-4} L

26. (a) 2.5×10^5 tons
(b) 9.9×10^{-4} g
(c) 1.50×10^{-8} L
(d) 3.00×10^5 bucks

27. (a) $c = 2\pi r = 2\pi(1.5 \times 10^{11}\text{ m}) = 9.4 \times 10^{11}$ m
(b) $c/12 = (9.4 \times 10^{11}\text{ m})/12 = 7.8 \times 10^{10}$ m

28. (a) Speed = $d/t = 2\pi r/t = 2\pi(4000\text{ mi})/24\text{ h} =$ 1047 mi/h
(b) 1047 mi/h (1 h/3600 s) = 0.291 mi/s

29. (a) $A = \pi r^2 = \pi(3.50\text{ in.})^2 =$ 38.5 in²
(b) $\pi r^2 = \pi(7.00\text{ in.})^2 =$ 154 in²
(c) 154 in²/38.5 in² = 4.00 times

30. 495¢/38.5 in² = 12.9¢/ in²
1395¢/154 in² = 9.06¢/ in² Larger better buy

ANSWERS TO RELEVANCE QUESTIONS

1.4 Actually, not many. The length, mass (weight), and time standard units in the mks, cgs, and British systems, respectively, are meter, kilogram, second; centimeter, gram, second; and foot, pound, second. How many common items bought at grocery and drug stores use these units? Perhaps pounds or kilograms, for example, are used for flour and sugar, or there are foot-long hot dogs, or the dimensions of some items in centimeters. Most item quantities are given in submultiples or nonstandard units, primarily for weight (mass) and volume—ounce, gram, and milligram; and gallon, quart, pint, ounce (fluid), liter, and milliliter. Can you think of anything bought in units of seconds? Maybe the time on a pay telephone or a children's ride in a store.

1.5 The distance one *walked* could be described using only length units. However, *walking* as an action or motion involves a *time* rate of change of position, and both length and time are needed to describe this, as will be learned in Chapter 2.

1.6 (1) Gallons (volume purchased) and (2) price per gallon are both necessary to determine the cost and complete the transaction. An error in the number of gallons delivered would result in paying an incorrect higher or lower price. An error in the price per gallon would do likewise, depending on the direction of the error.

1.7 The national debt is on the order of \$5.0 trillion or $\$5.0 \times 10^{12}$, and
Individual share = $(\$5.0 \times 10^{12})/(2.50 \times 10^8\text{ persons})$
= $\$2.0 \times 10^4$/person or \$20,000/person

ANSWERS TO STUDY GUIDE QUIZ

Multiple-Choice Questions

1. c 2. b 3. c 4. a 5. b
6. d 7. c 8. a 9. a 10. a

Short-Answer Questions

1. (1) Read the problem and identify the principles, then write down the given quantities in symbol notation. This clarifies the information given and organizes it so that it can be applied to the problem solution process.

   (2) Determine what is wanted and write it down. This establishes clearly what must be calculated in the problem and establishes the units that must be used in the answer.

   (3) Survey the equations available and determine which ones must be used, then perform the calculations and determine the appropriate units and the number of significant figures in which your answer must be expressed. This gives the ability to express the final answer in the form and with the units most appropriate for the data given in the problem.

2. Sight, smell, taste, touch, hearing.

3. A fundamental quantity is a physical characteristic concerning the phenomena that must be used when discussing or describing any aspect of the physical sciences. There are four fundamental quantities defined in this course: length, mass, time, and electric charge. A standard unit is a fixed and reproducible value estblished for the purpose of taking accurate measurements. A set of standard units is referred to as a system of units, the most widely used of which is the SI set of units based on the metric system. Standard units must be used as the basis for comparison when taking measurements of the fundamental quantities.

4. $100\ \cancel{\text{m}}\ \frac{1.09\ \text{yd}}{1\ \cancel{\text{m}}} = 109\ \text{yd}$, so the field would be 9 yd longer.

5. Density is equal to the mass of a sample of matter divided by the volume of that sample. Density is expressed in kg/m^3 when using SI units.

6. $23.8\ \cancel{\text{ft}}\ \frac{12\ \cancel{\text{in.}}}{1\ \cancel{\text{ft}}}\ \frac{2.54\ \text{cm}}{1\ \cancel{\text{in.}}} = 725\ \text{cm}$

   Since centi- means 1/100, this length will be: 725 cm (1/100) = 7.25 m

7. $\text{Mass} = 180\ \cancel{\text{lb}}\ \frac{1\ \text{kg}}{2.20\ \cancel{\text{lb}}} = 82\ \text{kg}$

   Since mass does not change when the man is transported to the Moon, his mass will be 82 kg in both places. However, his weight will change. His weight on the Moon would be only 1/6 that on Earth, or 30 lb.

8. $2{,}350{,}000 = 2.35 \times 10^6$

9. $6.66 \times 10^{10} = 66{,}600{,}000{,}000$

10. $25{,}000\ \cancel{\text{m}}\ \frac{1\ \text{km}}{1000\ \cancel{\text{m}}} = 25\ \text{km}$

Chapter 2

Motion

This chapter covers the basics of the description of motion. The concepts of position, speed, velocity, and acceleration are defined and physically interpreted, with applications to falling objects, circular motion, and projectiles. A distinction is made between average values and instantaneous values. Scalar and vector quantities are also discussed.

Problem solving is difficult for most students, primarily because they do not spend enough time practicing the problem-solving procedures given in Section 1.8. The authors have found it successful to assign a take-home quiz on several problems at the end of the chapter that must be handed in at the beginning of class. This may be followed by an in-class quiz on one of the take-home problems, for which the numerical values have been changed. This procedure provides the student with practice and helps gain confidence.

The student *Study Guide* that accompanies this textbook also provides detailed discussion and practice in the problem-solving process.

DEMONSTRATIONS

The linear air track may be used to demonstrate both velocity and acceleration. If an air track is not available, a 2-in. × 6-in. × 12-ft wooden plank may be substituted. It will be necessary to have a V groove cut into one edge of the plank to hold a steel ball of about 1-in. diameter. The ball will roll fairly freely in the V groove.

Also, various free-fall demonstrations are commercially available.

(General references to teaching aids are given in the Teaching Aids section at the back of this *Guide*.)

ANSWERS TO REVIEW QUESTIONS

1. b
2. A reference point or position. Reference points in Section 2.1 examples: intersection with traffic light, table, Georgia, origin.
3. The action of an object undergoing a continuous change in position.
4. d
5. d
6. a
7. A scalar quantity has magnitude only. A vector quantity has magnitude and direction.

8. Distance is the actual path length and is scalar. Displacement is the directed, straight-line distance between two points and is a vector. Distance is associated with speed, and displacement is associated with velocity.
9. (a) They are equal.
   (b) The average speed has a finite value, but the average velocity is zero because the displacement is zero.
10. Because of the small relative motions of stars. Most noticeable is the apparent motion of the Sun.
11. b
12. d
13. Either the magnitude or direction of the velocity, or both. An example of both is a child going down a slide at a playground.
14. Yes, the acceleration will merely change the instantaneous velocity.
15. A slight, but probably undetectable, difference due to air resistance.
16. Initial speed is zero. Initial acceleration is 9.8 m/s^2.
17. Yes. For an object thrown vertically upward, at its maximum height it is instantaneously at rest ($v = 0$), but it has an acceleration of g (otherwise it wouldn't fall back—can't turn off gravity).
18. c
19. Yes, in uniform circular motion.
20. Center-seeking.
21. (a) Spiral outward.
    (b) Fly off tangentially.
22. Yes, we are in circular motion in space. This can be sensed by the apparent motions of the Moon and of the Sun and other stars.
23. Inwardly toward the Earth's axis of rotation for (a) and (b). (c) None.
24. d
25. b
26. g and v_x
27. Initial velocity, projection angle, and air resistance.
28. No, it will always fall below a horizontal line because of the downward acceleration due to gravity.
29. No, the slower vertical motion at the top of the arc gives the illusion of "hanging" in the air.
30. Greater, so as to get more range or distance.
31. At an angle of less than 45°, so as to get a greater v_x.

ANSWERS TO CRITICAL THINKING QUESTIONS

1. More instantaneous. Think of having your speed measured by radar. This is an instantaneous measurement, and you get a ticket if you exceed the speed limit.
2. Not necessarily. The direction could be changing.
3. All are "accelerators" because they affect the magnitude or direction of the velocity.
4. So that a component of gravity provides some of the centripetal acceleration to negotiate the curve.
5. Yes. This is the case if the vertical components of the initial velocities are the same.

ANSWERS TO EXERCISES

1. $\bar{v} = d/t = 100 \text{ m}/15 \text{ s} = \underline{6.7 \text{ m/s}}$

2. $\bar{v} = 2\pi r = 2\pi(150 \text{ m})/[13 \text{ min }(60 \text{ s/min})] = \underline{1.2 \text{ m/s}}$

3. $t = d/v = (3.87 \times 10^8 \text{ m})/(3.00 \times 10^8 \text{ m/s}) = \underline{1.29 \text{ s}}$

4. $t = d/v = 630 \text{ mi}/(55.3 \text{ mi/h}) = \underline{11.4 \text{ h}}$

5. (a) $d_1 = vt = (45 \text{ mi/h})(2.0 \text{ h}) = \underline{90 \text{ mi}}$
   (b) $d_2 = (35 \text{ mi/h})(2.0 \text{ h}) = \underline{70 \text{ mi}}$
   (c) $\bar{v} = (90 \text{ mi} + 70 \text{ mi})/4.0 \text{ h} = \underline{40 \text{ mi/h}}$

6. (a) $\bar{v} = d/t = 180 \text{ km}/2.0 \text{ h} = \underline{90 \text{ km/h}}$
   (b) $\bar{v} = 180 \text{ km}/3.0 \text{ h} = \underline{60 \text{ km/h}}$
   (c) $\bar{v} = 360 \text{ km}/5.0 \text{ h} = \underline{72 \text{ km/h}}$

7. (a) $v = d/t = 300 \text{ km}/2.0 \text{ h} = \underline{150 \text{ km/h, east}}$
   (b) Same, since constant.

8. (a) $\bar{v} = d/t = 750 \text{ m}/20.0 \text{ s} = \underline{37.5 \text{ m/s, north}}$
   (b) Zero, since displacement is zero.

9. $\bar{a} = (v_f - v_o)/t = (12 \text{ m/s} - 0)/4.0 \text{ s} = \underline{3.0 \text{ m/s}^2}$

10. (a) $\bar{a} = (v_f - v_o)/t = (0 - 8.3 \text{ m/s})/1200 \text{ s} = \underline{-6.9 \times 10^{-3} \text{ m/s}^2}$
    (b) $\bar{v} = d/t = (5.0 \times 10^3 \text{ m})/(1.2 \times 10^3 \text{ s}) = \underline{4.2 \text{ m/s}}$
    (Needs to start slowing in plenty of time.)

11. (a) $\bar{a} = (v_f - v_o)/t = (12 \text{ m/s} - 0)/10 \text{ s} = \underline{1.2 \text{ m/s}^2 \text{ in direction of motion}}$
    (b) $\bar{a} = (18 \text{ m/s} - 0)/15 \text{ s} = \underline{1.2 \text{ m/s}^2 \text{ in direction of motion}}$

12. (a) $\bar{a} = (v_f - v_o)/t = (44 \text{ ft/s} - 0)/5.0 \text{ s} = \underline{8.8 \text{ ft/s}^2 \text{ in direction of motion}}$
    (b) $\bar{a} = (88 \text{ ft/s} - 0)/9.0 \text{ s} = \underline{9.8 \text{ ft/s}^2 \text{ in direction of motion}}$
    (c) $(66 \text{ ft/s} - 88 \text{ ft/s})/3.0 \text{ s} = \underline{-7.3 \text{ ft/s}^2 \text{ opposite direction of motion}}$
    (d) $\bar{a} = (66 \text{ ft/s} - 0)/12 \text{ s} = \underline{5.5 \text{ ft/s}^2 \text{ in direction of motion}}$

13. (a) $v = v_o + at = 0 + (2.0 \text{ m/s}^2)(5.0 \text{ s}) = \underline{10 \text{ m/s}}$
    (b) $d = \frac{1}{2}at^2 = \frac{1}{2}(2.0 \text{ m/s}^2)(5.0 \text{ s})^2 = \underline{25 \text{ m}}$

14. (a) $v = v_o + at$ or $60 \text{ m/s} = v_o + (40 \text{ m/s})(12 \text{ s}) = v_o + 48 \text{ m/s}$, so $v_o \neq 0$.
    (b) Had initial velocity of $v_o = 60 \text{ m/s} - 48 \text{ m/s} = 12 \text{ m/s}$ in direction of acceleration, or an initial velocity of $v_o = 60 \text{ m/s} + 48 \text{ m/s} = 108 \text{ m/s}$ in direction opposite acceleration.

15. (a) $v = v_o + gt = 0 + (9.80 \text{ m/s}^2)(1.00 \text{ s}) = \underline{9.80 \text{ m/s}}$, $a = g = \underline{9.80 \text{ m/s}^2}$
    (b) $v = (9.80 \text{ m/s}^2)(2.00 \text{ s}) = 19.6 \text{ m/s}$, $a = g = \underline{9.80 \text{ m/s}^2}$

16. $d_1 = \frac{1}{2}gt^2 = \frac{1}{2}(9.80 \text{ m/s}^2)(1.00 \text{ s})^2 = \underline{4.90 \text{ m}}$
    $d_2 = \frac{1}{2}(9.80 \text{ m/s}^2)(2.00 \text{ s})^2 = \underline{19.6 \text{ m}}$

17. (a) $v = v_o + gt = 0 + (32 \text{ ft/s}^2)(3.0 \text{ s}) = \underline{96 \text{ ft/s}}$ $[(0.682 \text{ mi/h})/(\text{ft/s})] = \underline{65 \text{ mi/h}}$
    (Pretty dangerous.)
    (b) $d = \frac{1}{2}gt^2 = \frac{1}{2}(9.8 \text{ m/s}^2)(3.0 \text{ s})^2 = \underline{44 \text{ m}}$
    $44 \text{ m }(32 \text{ floors}/100 \text{ m}) = \underline{14 \text{ floors}}$

18. $t = \sqrt{2d/g} = \sqrt{2(100 \text{ m})/(9.80 \text{ m/s}^2)} = \underline{4.52 \text{ s}}$

19. (a) $a_c = v^2/r = (10\text{ m/s})^2/70\text{ m} =$ 1.4 m/s² toward center
    (b) $a_c/g = (1.4\text{ m/s}^2)/)9.8\text{ m/s}^2) =$ 0.14 or 14%

20. $a_c = v^2/r = (25\text{ m/s})^2/500\text{ m} =$ 1.25 m/s² toward center

21. 0.55 s since same vertical motion.

22. (a) 200 m/s, west
    (b) 9.8 m/s², downward
    (c) $t = \sqrt{2y/g} = \sqrt{2(1000\text{ m})/(980\text{ m/s}^2)} = 14.3$ s, and
    $d = v_x t = (200\text{ m/s})(14.3\text{ s}) =$ 2860 m in front

ANSWERS TO RELEVANCE QUESTIONS

2.1 In terms of distance traveled and the time taken to travel the distance.

2.2 Probably speed. We commonly are concerned with how fast something is going, without regard to direction.

2.3 Many—every time you change speed and/or direction.

2.4 On the rotating Earth, gravity supplies the centripetal acceleration. Other examples might include going around a curve or circular track in a car, riding on a rotating amusement or playground ride—any time you are traveling in a circular path.

ANSWERS TO STUDY GUIDE QUIZ

Multiple-Choice Questions

1. c	2. b	3. c	4. b	5. b
6. a	7. c	8. c	9. b	10. a

Short-Answer Questions

1. Speed is defined as the distance (actual path traveled) divided by the time and is a scalar quantity. Velocity is defined as the displacement (straight path between starting and ending location) divided by the time and is a vector quantity.

2. A person walking at a constant rate in a straight line will have the same velocity at all points along her path of travel, and so her instantaneous velocity and her average velocity would be the same.

   A car that is slowing down as its brakes are applied would have a changing velocity, and so its instantaneous velocity when the brakes are first applied will not be the same as the average velocity of the car over the entire distance it takes to stop its forward motion.

3. (a) The magnitude of the velocity can change; that is, the object can speed up or slow down.
   (b) The direction of the velocity can change, as it does when an object is following a curved path (at a constant speed).

4. The magnitude of the velocity of a freely falling object will change if the object is moving straight up or straight down. Both the magnitude and the direction of the object's velocity will change if the object is in projectile motion and is following a two-dimensional parabolic path.

5. A baseball that is thrown horizontally near Earth's surface will be affected by gravity and will follow a parabolic path. This means that its horizontal velocity will remain constant but its vertical

5. A baseball that is thrown horizontally near Earth's surface will be affected by gravity and will follow a parabolic path. This means that its horizontal velocity will remain constant but its vertical velocity will increase as time progresses, which results in the curved (parabolic) path that the ball follows.

6. $v_f = v_o + \underline{at} = -2.7\text{ m/s} + (-9.8\text{ m/s})(3.5\text{ s})$

$= -2.7\text{ m/s} + (-34.3\text{ m/s}) = \underline{-37.0\text{ m/s (downward)}}$

7. $d = \underline{vt} = 7.5\text{ m/s}(10.0\text{ s}) = \underline{75\text{ m}}$

Note: No direction is required, as this is a scalar result. Also, the distance traveled is not dependent on the diameter of the track. It is not unusual to have more data available than are needed, so you must be able to pick out the parts that are relevant and use them to solve the problem.

8. Centripetal acceleration always points in the direction of the center of the circle that defines the curvature of the path being followed. This is often referred to as a radial direction, that is, along the radius of the circle and toward its center.

9. $a_c = v^2/r = (84\text{ m/s})^2 / (240\text{ m}) = 29.4\text{ m/s}^2 = 29\text{ m/s}^2$ (toward the center of the circle)

10. Air friction will slow down the motion of an object in projectile motion, causing it to have a lower maximum height and a shorter range than it would have had if air resistance were negligible.

Chapter 3

Force and Motion

This chapter is one of the most important in the textbook because it deals with Newton's laws of motion and gravitation, as well as the concepts of linear and angular momentum. The material naturally follows that of Chapter 2. With the foundations of kinematics established, the agents that produce motion are considered. This branch of mechanics is known as dynamics. Sufficient time should be spent on this material to be sure students have a firm understanding of the concepts.

Require students to make complete statements of Newton's laws and to give examples. When stating Newton's second law of motion, stress that the force is the unbalanced force acting on the total mass, that the mass is the total mass being accelerated, and that the acceleration is always in the direction of the unbalanced or net force. Acceleration is evidence of the action of an unbalanced force.

DEMONSTRATIONS

The linear air track can be used to illustrate Newton's laws of motion and the concept of impulse and momentum.

The Atwood machine is an excellent demonstration for illustrating Newton's second law of motion. Best results can be obtained if the student is led through the demonstration (see the *Laboratory Guide*) by questions rather than having the instructor merely perform the experiment.

An apple may be brought to class to illustrate the idea of one newton of weight. Be sure the apple weighs about 3.6 ounces. Also, a spring balance calibrated in newtons can be displayed supporting a 1-kg mass.

Free-fall can be demonstrated with a feather and a coin in a glass tube from which the air can be evacuated. Let your students handle the glass tube for the best results.

Newton's third law of motion can be demonstrated by using the toy rocket that holds water under pressure—equal and opposite forces are demonstrated as the rocket accelerates along a string. Releasing a blown balloon also illustrates the law.

The law of conservation of angular momentum is demonstrated dramatically using a turntable or rotating stool and two masses (e.g., 1 kg each) held in the hands. While rotating, the masses are brought closer to the body (reduced moment of inertia), the rate of rotation increases. When beginning the demonstration, point out that you can't get started by yourself. You must have an external force or torque, which can be supplied by a student. Students will often ask to try the demonstration. Permission should be granted with caution. The rotation can make a person quite dizzy.

(General references to teaching aids are given in the Teaching Aids section at the back of this *Guide*.)

ANSWERS TO REVIEW QUESTIONS

1. a
2. c
3. Usually that the person is sluggish, and slow to move.
4. Dishes and glasses remain in place because of inertia.
5. Fuller roll has more mass, and hence more inertia.
6. To provide an external force on stopping so that persons will not continue in motion.
7. b
8. d
9. (a) $F \propto a$
   (b) $a \propto 1/m$
10. (a) Yes, if the sum of the forces is zero.
    (b) Yes, uniform motion (constant velocity).
11. Zero, since velocity is constant.
12. (a) A unit of force
    (b) 8 N.
13. (a) Ten times the force, but also ten times the inertia; therefore, it falls at the same rate.
    (b) Similar, except that the acceleration of the rocks would be less—that is, $g/6$.
14. c
15. a
16. The distance between two point masses.
17. Because $F \propto 1/r^2$, and F approaches zero only as r approaches infinity.
18. The variation is small over the range of altitudes of common activities.
19. (a) No. Gravity supplies the centripetal force, and by definition, there is weight.
    (b) Only if the resultant of two or more gravitational forces on an object is zero.
20. b
21. c
22. No, the pad is there only to hold the rocket. The expanding combustion gases exert a force on the rocket, and the rocket exerts a force on the gases. Consider firing a rocket in space—there is nothing to "push against" there.
23. The two forces of the force pair act on different objects.
24. Yes. There is a net force on the accelerating object, which exerts a reaction force(s) on another object(s).
25. Equal and opposite forces on the gun and bullet. Because the gun is more massive, it has a smaller acceleration, but enough to give a "kick."
26. d
27. c
28. Has or loses "drive."
29. In the absence of an unbalanced external force, an object or system has a constant velocity and hence a constant momentum.

30. Both (a) and (b) increase the contact time so as to decrease the impulse force that could cause breakage or injury.
31. The rotational speed of the key increases because of the conservation of momentum—smaller radius, greater speed.
32. Through the conservation of angular momentum. Tucking reduces the r of the mass distribution and the rotational speed increases.
33. No. The helicopter body would rotate opposite to the rotor to conserve angular momentum.

ANSWERS TO CRITICAL THINKING QUESTIONS

1. The first law is consistent with the second law, since in the absence of a force, the acceleration is zero and the velocity is either zero or constant. The second law is implied in the first, but no relationships of mass and acceleration are given.
2. Considering only the gravitational attractions of the Earth and Moon, at some point in space the forces would cancel and there would essentially be no weight, by definition. After this point was passed, there would be a net force and an acceleration toward the Moon.
3. It would accelerate toward the center and pass through. On going toward the other side, it would slow down as a result of gravitational attraction (central mass), eventually stop, and start to fall toward the original side. This would repeat, and the object would oscillate back and forth in the hole.
4. The adhesion between the water and the clothes is not sufficient to provide the necessary centripetal force for the water to rotate with the clothes, and hence the water becomes separated.
5. Players floating around trying to get above the basket to throw or slam dunk the ball through the hoop.

ANSWERS TO EXERCISES

1. $F = ma = (4.0\text{ kg})(6.0\text{ m/s}^2) = \underline{24\text{ N}}$
2. $a = F/m = 2.1\text{ N}/(7.0 \times 10^{-3}\text{ kg}) = \underline{300\text{ m/s}^2}$
3. $a = F/m = 850\text{ N}/1000\text{ kg} = \underline{0.85\text{ m/s}^2}$
4. $m = F/a = 1500\text{ N}/2.5\text{ m/s}^2 = \underline{600\text{ kg}}$
5. $w = mg = (6.0\text{ kg})(9.8\text{ m/s}^2) = \underline{59\text{ N}}$
6. 59 N (same as in Exercise 5)
7. (a) 140 lb (4.45 N/lb) = $\underline{623\text{ N}}$
   (b) Personal
8. $w = mg = (0.500\text{ kg})(9.80\text{ m/s}^2) = \underline{4.90\text{ N}}$
9. $T = mg = (5.0\text{ kg})(9.8\text{ m/s}^2) = \underline{49\text{ N}}$
10. (a) $T_1 = (m_1 + m_2)g = (6.0\text{ kg})(9.8\text{ m/s}^2) = \underline{59\text{ N}}$
    (b) $T_2 = m_2 g = (2.0\text{ kg})(9.8\text{ m/s}^2) = \underline{20\text{ N}}$

11. (a) $a = F/(m_1 + m_2) = 10\text{ N}/5.0\text{ kg} = \underline{2.0\text{ m/s}^2}$, in the direction of the applied force.
    (b) $a = (F - f)/(m_1 + m_2) = 6.0\text{ N}/5.0\text{ kg} = \underline{1.2\text{ m/s}^2}$, in the direction of the applied force.

12. $a_1 = F/m_1 = 12\text{ N}/4.0\text{ kg} = \underline{3.0\text{ m/s}^2}$
    $a_2 = F/(m_1 + m_2) = 12\text{ N}/6.0\text{ kg} = \underline{2.0\text{ m/s}^2}$
    $(3.0\text{ m/s}^2 - 2.0\text{ m/s}^2)/(3.0\text{ m/s}^2) \times 100\% = \underline{33\%\text{ decrease}}$

13. $F = Gm_1m_2/r^2 = (6.67 \times 10^{-11}\text{ N-m}^2/\text{kg}^2)(1.5\text{ kg})(1.5\text{ kg})/(0.35\text{ m})^2$
    $= \underline{1.2 \times 10^{-9}\text{ N}}$

14. (a) $F = Gm_1m_2/r^2 = (6.67 \times 10^{-11}\text{ N-m}^2/\text{kg}^2)(10^3\text{ kg})(10^3\text{ kg})/(50\text{ m})^2$
    $= \underline{2.7 \times 10^{-8}\text{ N}}$
    (b) Much, much smaller; $w = mg = (10^3\text{ kg})(9.8\text{ m/s}^2) = \underline{9.8 \times 10^3\text{ N}}$

15. (a) $r_2 = 3r_1$, and $F_2/F_1 = (r_1/r_2)^2 = (1/3)^2 = \underline{1/9}$
    (b) $r_2 = r_1/2$, and $F_2/F_1 = (2)^2 = \underline{4}$

16. (a) $r_2 = 2r_1/3$, and $F_2/F_1 = (r_1/r_2)^2 = (3/2)^2 = \underline{9/4 = 2.25}$
    (b) $r_2 = 5r_1$, and $F_2/F_1 = (1/5)^2 = \underline{1/25}$

17. $g = GM_E/R_E^2 = (6.67 \times 10^{-11}\text{ N-m}^2/\text{kg}^2)(5.96 \times 10^{24}\text{ kg})/(6.37 \times 10^6\text{ m})^2$
    $= \underline{9.80\text{ m/s}^2}$

18. $g_M/g_E = (M_M/M_E)(R_E/R_M)^2$
    $= [(7.35 \times 10^{22}\text{ kg})/(5.96 \times 10^{24}\text{ kg})][(6.37 \times 10^6\text{ m})/(1.74 \times 10^6\text{ m})]^2$
    $= \underline{0.165 \approx 1/6}$ $(1/6 = 0.167)$

19. $g/g_E = (R_E/R_E + h)^2 = [(6.37 \times 10^6\text{ m})/(7.37 \times 10^6\text{ m})]^2 = \underline{0.747}$
    $g = (0.747)g_E = (0.747)(9.80\text{ m/s}^2) = \underline{7.32\text{ m/s}^2}$

20. $g/g_E = (R_E/R_E + h)^2 = \frac{1}{2}$, and $R_E/(R_E + h) = \underline{0.707}$
    $h = 0.293R_E/0.707 = 0.293(6.37 \times 10^6\text{ m})/0.707 = \underline{2.64 \times 10^6\text{ m}}$

21. (a) $w_M = w_E/6 = 180\text{ lb}/6 = \underline{30\text{ lb}}$,
    (b) Personal

22. (a) $w_E = 6w_M = 6(18\text{ N}) = \underline{108\text{ N}}$
    (b) $m = 108\text{ N}/g = \underline{11\text{ kg}}$
    (c) $\underline{11\text{ kg}}$

23. $p = mv = (1.3 \times 10^4\text{ kg})(20\text{ m/s}) = \underline{2.6 \times 10^5\text{ kg-m/s, east}}$

24. $p = mv = (9.0 \times 10^2\text{ kg})(30\text{ m/s}) = \underline{2.7 \times 10^4\text{ kg-m/s, north}}$
    Less, about 10 times less.

25. $\bar{F}\,\Delta t = \Delta p = \underline{2.2\text{ N–s or }2.2\text{ kg–m/s}}$

26. (a) $\bar{F}\,\Delta t = \Delta p = mv = (5.0 \times 10^{-2}\text{ kg})(30\text{ m/s}) = \underline{1.5\text{ kg-m/s}}$
    (b) $\bar{F}\,\Delta p/\Delta t = (1.5\text{ kg-m/s})/0.010\text{ s} = \underline{1.5 \times 10^2\text{ N}}$

27. $v_2 = r_1v_1/r_2 = (600 \times 10^6\text{ mi})(15{,}000\text{ mi/h})/(100 \times 10^6\text{ mi}) = \underline{90{,}000\text{ mi/h}}$

28. $v_2 = r_1v_1/r_2 = (2.0 \times 10^8\text{ km})(1.2 \times 10^6\text{ m/s})/(8.0 \times 10^8\text{ km}) = \underline{3.0 \times 10^5\text{ m/s}}$

ANSWERS TO RELEVANCE QUESTIONS

3.1. Advantage: when you want something to remain in motion or at rest—for example, a car coasting a greater distance, or pulling (jerking) a magazine from a stack without toppling the stack, respectively. Disadvantage: when you want to stop something, or set something in motion—for example, trying to block a running fullback, or tearing a paper towel from a small roll, respectively.

3.2 Use known weight in pounds and convert to newtons (1 lb = 4.45 N).

3.4 Gravity (downward) and the force on you by the Earth (upward)—equal and opposite. Reaction forces are the gravitational attraction by you on the Earth (upward) and the force you exert on the Earth (downward)—equal and opposite.

3.5 Yes, a force and a lever arm, and so a torque (otherwise you couldn't open the door). There is a change in momentum because there is a torque, and so momentum is not conserved.

ANSWERS TO STUDY GUIDE QUIZ

Multiple-Choice Questions

1. c 2. a 3. c 4. d 5. b
6. b 7. a 8. d 9. b 10. b

Short-Answer Questions

1. Newton's first law of motion states that an object will remain at rest or in uniform motion in a straight line unless acted upon by an external, unbalanced force.
2. Inertia is the tendency of an object to resist any change in its state of motion. Inertia is related to the mass of an object; that is, the more massive the object, the more it resists an acceleration or deceleration when an external force is applied to it.
3. If two boys are pushing on opposite sides of a large crate that is resting on a slippery surface such as an icy driveway, but one is pushing much harder than the other, the crate will experience an unbalanced force that will result in the acceleration of the crate in the direction in which the stronger force is being applied.
4. Mass is an intrinsic property determined by the amount of matter present in any body. This quantity remains the same no matter where the body is located in the universe. The weight of the body, on the other hand, is the force with which the body is attracted to Earth, or to any other large celestial object near which the body is located. Near Earth's surface, the weight of any body is equal to its mass multiplied by the acceleration of gravity (g).
5. The weight of a person located on the surface of Earth is directly proportional to the mass of Earth and inversely proportional to the radius of Earth squared. If the same object is located on the Moon, its weight will be directly proportional to the mass of the Moon and inversely proportional to the radius of the Moon squared. Since the masses and the radii of Earth and the Moon are quite different, the value of the person's weight will not be the same in both places.
6. Momentum, the product of an object's mass multiplied by its velocity, will stay the same from one instant to the next unless an external unbalanced force is applied to the object. In collision processes the interaction between objects quite often does not involve the application of any outside unbalanced forces, and so the total momentum of the system is conserved; that is, it remains the same after the collision as it was before the collision took place.

7. Here the mass can be found by applying Newton's second law. If we solve for the mass, we get $m = F/a$, so $m = 80\ \text{N}/4.0\ \text{m/s}^2 = 20\ \text{kg}$.
8. $W = mg = 20\ \text{kg}(9.8\ \text{m/s}^2) = 200\ \text{N}$
9. $p = mv = 6000\ \text{kg}(0.25\ \text{m/s}) = 150\ \text{kg–m/s}$ due east. Note that since momentum is a vector quantity, the direction must be stated and will be the same as that of the velocity of the bulldozer, in this case due east.
10. Here we can use the relatonship between the applied impulse and the change in momentum of the bulldozer to find the force.

$$F_t = mv_f - mv_i, \text{ where } v_f = 0 \text{ (it will be at rest) and } v_i = 0.25\ \text{m/s}$$

$$F = (mv_f - mv_i)/t = (6000\ \text{kg} \times 0\ \text{m/s} - 6000\ \text{kg} \times 0.25\ \text{m/s})/32\ \text{s}$$

$$F = (-1500\ \text{kg-m/s})/32\ \text{s} = -47\ \text{N to the west}$$

The – sign means that the force is in the opposite direction to the original velocity, which was to the east, and so the force that decelerates the bulldozer must be applied toward the west.

Chapter 4

Work and Energy

This chapter should be covered thoroughly in lecture and assignment. The relationship of work and energy is of the utmost importance in understanding many daily activities. Also, the development of the physical environment is closely associated with the control of energy. The law of conservation of energy is one of the most important general laws, and has been a key to many of nature's secrets. Thus, it is important for the student to know the meanings of work and energy, and to be familiar with various forms of energy.

Although this chapter deals primarily with general concepts and mechanical energy, it should be pointed out how easy it is to change other types of energy, such as chemical and electrical energy to other forms and use it to do work.

The textbook tries to get the student to think in terms of symbols, and this is a good chapter to stress this kind of thinking. For example: When referring to kinetic energy, think $\frac{1}{2}mv^2$, and when thinking of gravitational potential energy, think *mgh.*

DEMONSTRATIONS

A simple pendulum can be used to display the transformation of potential energy to kinetic energy and vice versa. As the pendulum swings back and forth, ask the students at what stages the velocity, acceleration, potential energy, and kinetic energy have their minimum and maximum values.

Demonstrate the examples of work as shown in the illustrations in the textbook.

A radiometer can be used to show that light can do work.

(General references to teaching aids are given in the Teaching Aids section at the back of this *Guide*.)

ANSWERS TO REVIEW QUESTIONS

1. d
2. b
3. A force acting through a parallel distance.
4. No, there must be motion. No work is done in holding an object stationary, but work is done in lifting.
5. (a) $W = Fd$, therefore N-m or joule (J).
   (b) N-m = (kg-m/s^2)-m = kg-m^2/s^2.
6. No work while stationary. Work was done on the weights in lifting.

7. b
8. c
9. When something has energy, it possesses the capability to do work.
10. Braking distance is directly proportional to the square of the velocity, $Fd = \frac{1}{2}mv^2$, and $d \propto v^2$.
11. Yes, if the zero reference point is chosen at the height of the book.
12. (a) The height depends on the initial kinetic energy, $mgh = \frac{1}{2}mv^2$.
    (b) By the conservation of energy, it would have the same speed as it had initially.
13. Yes, relocate the arbitrary zero reference position.
14. For energy conversion, kinetic energy into potential energy and height. Long jumpers want a large horizontal velocity so as to get distance.
15. d
16. c
17. Total energy includes all forms of energy. Mechanical energy is the sum of the kinetic and potential energies.
18. A system is something enclosed within boundaries, which may be real or imaginary.
19. When energy neither enters or leaves a system and thus has a constant value.
20. (a) a and e. (b) c. (c) c. (d) a and e. (e) a and e. (f) c. (g) a and e. (h) c. (i) c. (j) a and e.
21. Same initial speed from same height. Both will have same speed on striking the ground. Conservation of energy.
22. (a) b and c. (b) a. (c) a. (d) b and c. (e) b and c. (f) a. (g) b, c; and e, d when going toward $h = 0$. (h) a; and e, d, when spring is compressing. (i) a; and e, d, when spring is compressing. (j) b, c; and e, d when going toward $h = 0$.
23. c
24. a
25. (a) $P = W/t$, so J/s or watt (W).
    (b) J/s = N-m/s = (kg-m/s^2)-m/s = kg-m^2/s^3.
26. $P = W/t$, so person A, with the shorter time, expends more power.
27. (a) More done in a given time.
    (b) Doing a given amount of work faster.
28. Kilowatt-hour (kWh), a unit of energy.
29. Stove, air conditioner, and hot water heater.
30. b
31. a
32. 100 J/s $\times$ 3600 s = 3.6×10^5 J
33. Radiant, chemical, nuclear, sound, and heat.
34. See discussion in Section 4.5.

ANSWERS TO CRITICAL THINKING QUESTIONS

1. Yes, a bucket of fuel or anything flammable contains intrinsic energy. So does a bucket of hot water or certain nuclear materials (see Chapter 10).
2. Yes, but not through smell. We can feel heat and vibrations, see energetic phenomena and light, and hear sound. It is doubtful that you could taste energy.
3. Piecework involves power because the more energy expended per unit time, the greater the output and the more pay. An hourly rate implies a more constant power output, or at least compensation for same.
4. Yes, because solar energy powers the hydrologic cycle that brings moisture inland from the oceans.
5. For such a machine, you would have to have the same output as input (no loss), which requires a 100% efficiency. If the efficiency were greater than 100%, you would get more energy (or work) out than you put in—a creation of energy.

ANSWERS TO EXERCISES

1. $W = Fd = (20\ \text{N})(0.50\ \text{m}) = \underline{10\ \text{J}}$
2. $F = W/d = 300\ \text{J}/2.0\ \text{m} = \underline{1.5 \times 10^2\ \text{N}}$
3. $W = mgh = (5.0\ \text{kg})(9.8\ \text{m/s}^2)(0.45\ \text{m}) = \underline{22\ \text{J}}$
4. $W = mgh = (4.0\ \text{kg})(9.8\ \text{m/s}^2)(2.0\ \text{m}) = \underline{78\ \text{J}}$
5. $W = Fd = (0.60)(200\ \text{N})(6.0\ \text{m}) = \underline{7.2 \times 10^2\ \text{J}}$
6. $W = F_1h + F_2d = (100\ \text{N})(0.50\ \text{m}) + (75\ \text{N})(4.0\ \text{m}) = \underline{3.5 \times 10^2\ \text{J}}$
7. $E_k = \frac{1}{2}mv^2 = \frac{1}{2}(1000\ \text{kg})(25\ \text{m/s})^2 = \underline{3.1 \times 10^5\ \text{J}}$
8. $W = E_k = 3.1 \times 10^5\ \text{J}$ (from Exercise 7)
9. $E_k = \frac{1}{2}mv^2 = \frac{1}{2}(20\ \text{kg})(9.0\ \text{m/s})^2 = \underline{8.1 \times 10^2\ \text{J}}$
10. $E_{kb} = \frac{1}{2}mv^2 = \frac{1}{2}(2.0 \times 10^{-3}\ \text{kg})(4.0 \times 10^2\ \text{m/s})^2 = \underline{1.6 \times 10^2\ \text{J}}$
    $E_{ko} = \frac{1}{2}mv^2 = \frac{1}{2}(6.4 \times 10^7\ \text{kg})(10\ \text{m/s})^2 = \underline{3.2 \times 10^9\ \text{J}}$
    Ocean liner has greater kinetic energy.
11. Need to compute E_k for each speed.
    $\Delta E = E_{k2} - E_{k1} = \frac{1}{2}m(v^2{}_2 - v^2{}_1) = \frac{1}{2}(1.0\ \text{kg})[(6.0\ \text{m/s})^2 - (2.0\ \text{m/s})^2] = \underline{16\ \text{J}}$
12. $W = \Delta E = \frac{1}{2}m(v^2{}_2 - v^2{}_1) = \frac{1}{2}(10^3\ \text{kg})[(8.3\ \text{m/s})^2 - (25\ \text{m/s})^2] = \underline{-2.8 \times 10^5\ \text{J}}$
13. (a) $E_{k2}/E_{k1} = (v_2/v_1)^2 = [(3.0\ \text{m/s})/(1.0\ \text{m/s})]^2 = \underline{9}$
    (b) $(v_2/v_1)^2 = [(8.0\ \text{m/s})/(2.0\ \text{m/s})]^2 = \underline{16}$
14. (a) $E_{k2}/E_{k1} = (v_2/v_1)^2 = [(40\ \text{km/h})/(30\ \text{km/h})]^2 = \underline{1.8}$
    (b) $(v_2/v_1)^2 = [(30\ \text{km/h})/(40\ \text{km/h})]^2 = \underline{0.56}$
15. $E_p = mgh = (3.00\ \text{kg})(9.80\ \text{m/s}^2)(-10.0\ \text{m}) = \underline{-294\ \text{J}}$
    Below zero point.

16. E_p = $\underline{+294\text{ J}}$ (same magnitude as in Exercise 15)

17. $\frac{1}{2}E_p$ (lost) $= \frac{1}{2}mgh = \frac{1}{2}mg(10\text{ m})$, and $h = \underline{5.0\text{ m}}$

18. $(0.33)E_p = (0.33)mgh = (0.33)mg(0.60\text{ m})$, and $h = \underline{2.0\text{ m}}$

19. (a) $E = mgh = (50\text{ kg})(9.8\text{ m/s}^2)(10\text{ m}) = \underline{4.9 \times 10^3\text{ J}}$,
    (b) Same

20. $mgh = \frac{1}{2}mv^2$, and $v = \sqrt{2gh}$ (h from top)
    (a) $v = \sqrt{2g(10.0\text{ m})} = \underline{14.0\text{ m/s}}$
    (b) $v = \sqrt{2g(15.0\text{ m})} = \underline{17.1\text{ m/s}}$

21. (a) $v = \sqrt{2gh} = \sqrt{2g(20\text{ m})} = \underline{20\text{ m/s}}$
    (b) $v = \sqrt{2g(15\text{ m})} = \underline{17\text{ m/s}}$

22. $E = mgh = (1.00\text{ kg})(9.80\text{ m/s}^2)(50.0\text{ m}) = 490\text{ J}$
    (a) $v = gt = g(1.00\text{ s}) = 9.80\text{ m/s}$
    $E_k = \frac{1}{2}mv^2 = \frac{1}{2}(1.00\text{ kg})(9.80\text{ m/s})^2 = \underline{48.0\text{ J}}$
    $E_p = 490\text{ J} - 48.0\text{ J} = \underline{442\text{ J}}$
    (b) $v = gt = g(2.00\text{ s}) = \underline{19.6\text{ m/s}}$
    $E_k = \frac{1}{2}(1.00\text{ kg})(19.6\text{ m/s})^2 = \underline{192\text{ J}}$
    $E = 490\text{ J} - 192\text{ J} = \underline{298\text{ J}}$

23. $P = W/t = 7.2 \times 10^2\text{ J/30 s} = \underline{24\text{ W}}$

24. $P = W/t = 3.5 \times 10^2\text{ J/10 s} = \underline{35\text{ W}}$

25. (a) $W = Fh = (500\text{ N})(4.0\text{ m}) = \underline{2.0 \times 10^3\text{ J}}$
    (b) $P = W/t = 2.0 \times 10^3\text{ J/25 s} = \underline{80\text{ W}}$

26. $P = W/t = mgh/t = 154\text{ lb}(1\text{ kg/2.2 lb})(9.8\text{ m/s}^2)(6.0\text{ m})/5.0\text{ s} = \underline{8.2 \times 10^2\text{ W}}$

27. $P = wh/t = (20\text{ lb})(22\text{ ft})/40\text{ s} = 11\text{ ft-lb/s}\ [(1\text{ hp}/550\text{ ft-lb/s})] = \underline{0.020\text{ hp}}$

28. $W = Pt = 3.0\text{ hp}[(550\text{ ft-lb/s/hp})](3600\text{ s}) = \underline{5.9 \times 10^6\text{ ft-lb}}$

29. $E = Pt = (1.65\text{ kW})(2/3\text{ h}) + (1.20\text{ kW})(3/4\text{ h}) + (1.25\text{ kW})(1/6\text{ h})$
    $+ (0.10\text{ kW})(6.0\text{ h}) = 2.81\text{ kWh}\ (\$0.12/\text{kWh}) = \underline{\$0.34}$

30. (a) $E = Pt = (1.25\text{ kW})(4.0/60\text{ h}) = \underline{0.083\text{ kWh}}$
    (b) 0.083 kWh (\$0.10/kWh) = $\underline{\$0.0083}$ (less than a penny)

ANSWERS TO RELEVANCE QUESTIONS

4.1 Work is done *against* gravity in lifting your mass up the stairs. Work is done *by* the muscles in your legs and feet.

4.2 Yes, this is numerically possible. If there is no motion, the kinetic energy is zero. If the student selects his or her position as the zero reference point, then the potential energy is zero.

4.4 The primary method is to reduce the kWh consumption through such things as turning off lights when they are not needed, turning off the TV when it is not being viewed, using lower thermostat settings in the winter and higher settings in the summer, and so on.

ANSWERS TO STUDY GUIDE QUIZ

Multiple-Choice Questions

1. c	2. a	3. b	4. d	5. a
6. b	7. b	8. d	9. a	10. a

Short-Answer Questions

1. (1) Work can be calculated as the change in total energy of an object, or (2) work can be calculated by multiplying the applied unbalanced force by the distance through which the force is applied.

2. Electrical *work* is measured in units of kilowatt-hours. Electrical power is measured in watts or kilowatts, and time can be measured in seconds or hours. The most common commercial unit for work is the derived unit combining these units, the kilowatt-hour, but the unit watt-second can also be used.

3. Any moving, massive object possesses *kinetic energy*. The kinetic energy of any object can be calculated using the equation $E_k = 1/2\ mv^2$.

4. In this case, the gravitational potential energy decreases, the kinetic energy of the falling object must increase in just such a way that the sum of the two will be constant throughout the fall. This requires that no work be done by friction, or any other force, on the falling object, or the total energy will not be conserved.

5. Since work is equal to the applied, unbalanced force multiplied by the parallel distance through which the force is applied, the work done by the man is

$$W = Fd = 820\text{ N}(3.5\text{ m}) = 2900\text{ J}$$

6. Electrical work is calculated by multiplying the electrical power by the time. In this case, the work is

$$W = Pt = 1500\text{ W}(4.5\text{ min})(60\text{ s}/1\text{ min}) = 410{,}000\text{ Ws}$$

or

$$= 1.5\text{ kW}(4.5\text{ min})(1\text{ h}/60\text{ min}) = 0.11\text{ kWh}$$

7. $E_p = mgh = 7.50\text{ kg}(9.80\text{ m/s}^2)(5 \times 3.20\text{ m}) = 1176\text{ J} = 11{,}800\text{ J}$

8. We will use conservation of energy, so the E_k of her running will be converted into E_p at the top of her swing.

$$mgh = 1/2\ mv^2 \quad \text{so} \quad h = v^2/2g \quad \text{(after the } m\text{'s are cancelled)}$$

$$h = (7.00\text{ m/s})^2/[2(9.80\text{ m/s}^2)] = 2.50\text{ m}$$

9. If we find the total energy at the top and the E_p at the new height, the difference between them will be the E_k at the new height.

Top energy at top = $E_p = mgh = 0.035\text{ kg}(9.8\text{ m/s}^2)(14.0\text{ m}) = 4.8\text{ J}$
E_p at 6 m = $0.035\text{ kg}(9.8\text{ m/s}^2)(6.00\text{ m}) = 2.1\text{ J}$
E_k at 6 m = $E_T - E_p = 4.8\text{ J} - 2.1\text{ J} = 2.7\text{ J}$

An alternative method would be to calculate the velocity of the brush after it had fallen to 6.00 m above the ground, and then use $E_k = 1/2\ mv^2$.

10. Again, conservation of energy can be used.

$E_k = E_t - E_p$, but E_p at the bottom of the hill is 0.

$E_k = mgh = 2200\ \text{kg}(9.80\ \text{m/s}^2)(9.00\ \text{m}) = 194{,}000\ \text{J}$

Then, since $E_k = \frac{1}{2}mv^2$,

$$v = \sqrt{2E_k/m} = \sqrt{2(194{,}000\ \text{J})/2200\ \text{kg}} = \sqrt{176\ \text{m}^2/\text{s}^2} = 13.3\ \text{m/s}$$

Chapter 5

Temperature and Heat

Chapter 5 is an important chapter because temperature and heat are two of the most common physical concepts the student experiences. However, its content is not referred to in great detail in the succeeding chapters on atomic theory and chemistry. In general, temperature measurements are given, and we say that heat is a form of energy. Hence, it is important that a basic understanding of temperature and heat is obtained here.

The instructor may wish to expand on the microscopic basis of heat. For the most part, the chapter is concerned with the measurement of macroscopic quantities of heat, such as specific heat and latent heat. The general trend is to express these heats in joules (J). However, calculations will be done primarily in kilocalories (kcal) because of the difficulty of adding numbers expressed in powers of 10 which is necessary when using joules. Calculations are much easier when done in kilocalories, and the results can be converted to joules if so desired.

Because heat transfer has many applications in daily life, this is an important and interesting topic that should be covered in some detail.

DEMONSTRATIONS

Crooke's thermometer demonstrates radiant heat.
Franklin's palm glass demonstrates transfer of heat.
Ring and ball demonstrates expansion of metals (and holes in metals).
Compound bar demonstrates unequal expansion of metals.

A thermometer may be calibrated in class by using boiling water for the steam point and ice water for the ice point. Uncalibrated thermometers are available, and students find it interesting and obtain a grasp for temperature scales when the interval between the ice and steam points is divided into degrees. (How should it be done?) Also, have a calibrated thermometer on hand so that you can check and see how accurate your calibration is.

The bulb of one of two thermometers may be painted black and exposed to sunlight or a heat lamp to show the difference in radiation absorption.

General references to teaching aids are given in the Teaching Aids section at the back of this *Guide*.

ANSWERS TO REVIEW QUESTIONS

1. d
2. b
3. (a) Because it gives only indications of hotness and coldness—that is, relative terms, rather than absolute energy values.
   (b) Thermal expansion of solids, liquids, and gases, which is calibrated with a temperature scale.
4. 32°F and 212°F; 0°C and 100°C; 273 K and 373 K
5. Because of the thermal expansion of the bimetallic coil on which it sits.
6. a
7. a
8. Heat added to a system becomes part of its internal energy. That is, heat is the energy transferred to or removed from a system.
9. A big Calorie (kcal) is 1000 cal, and a difference of 999 cal.
10. The British thermal unit (Btu); 1 Btu = 0.25 kcal = 250 cal.
11. To act as expansion joints and prevent cracking and damage from thermal expansion.
12. a
13. b
14. It is a measure of J/kg-C° for a particular substance and is characteristic of or specific for that substance.
15. Because it has a relatively high specific heat and can store more energy with less temperature increase.
16. Because water has a high specific heat and it takes some time to remove the intrinsic heat and lower the water temperature to the freezing point.
17. During a phase change, when it goes into breaking of bonds or the separation of molecules in changing the phase.
18. Specific heat, c = J/kg-C°; latent heat, J/kg. The latent heat process occurs at a particular temperature, hence there is no temperature change.
19. c
20. b
21. Heat is removed from the system (balloon), and negative work is done as the balloon collapses.
22. First law: Energy is conserved in thermodynamic processes. Second law: The direction of a process, and whether or not a process will take place spontaneously
23. Thermal efficiency: What work is actually done for a given heat input. Ideal efficiency: A maximum theoretical efficiency that neglects energy losses. As such, it sets an unattainable upper limit. The percentage of useful work obtained from heat input.
24. It goes into mechanical work done against friction, and is part of the thermal energy rejected to the cold-temperature reservoir. This heat is lost in cooling and condensing the steam after the expansion process that turns the turbine.
25. First law (conservation of energy) and second law (entropy increases in every natural process).
26. b
27. b

28. Thermal conductors: silver, copper, aluminum (metals). Thermal insulators: cloth, Styrofoam, wood. A difference in electron mobility and air space.
29. The material of the floor has greater thermal conductivity.
30. They do not conduct or convect because of poor conductivity and partial vacuum, respectively. A Thermos bottle also has a reflective coating on its walls to prevent radiation losses.
31. (a) No, inasmuch as a lot of heat goes up the chimney.
    (b) No, it provides insulating air spaces.
32. a
33. Temperature and pressure.
34. Solid: definite shape and volume. Liquid: definite volume, assumes shape of container. Gas: no definite shape or volume. (Volume may be restricted to a rigid container.)

ANSWERS TO CRITICAL THINKING QUESTIONS

1. Yes, this occurs at −40°. The Fahrenheit and Celsius degree intervals are different, which allows them to "catch up" with each other, so to speak.
2. Some cold air may be blown in, but in general, inside warm air is transferred outside because of the temperature difference.
3. The topping or filling has a large specific heat and does not cool as quickly.
4. The latent heat of a phase change depends on the molecular bonding of a substance and the work needed to effect the phase change.
5. First, second, and third laws of thermodynamics, respectively.

ANSWERS TO EXERCISES

1. $T_F = (9/5)T_C + 32 = (9/5)15° + 32 = \underline{59°F}$
2. $T_F = (9/5)180° + 32 = \underline{356°F}$
3. $T_C = 5/9(T_F - 32) = 5/9(68° - 32) = \underline{20°C}$
4. $T_C = 5/9(98.6° - 32) = \underline{37°C}$
5. (a) $T_C = 5/9(-40° - 32) = \underline{-40°C}$
   (b) $T_K = T_C + 273 = -40° + 273 = \underline{233\text{ K}}$
6. (a) $T_C = T_K - 273 = 3 - 273 = \underline{-270\text{ K}}$
   (b) $T_F = (9/5)(-270) + 32 = \underline{-454°F}$
7. $E - E_k = mgh - \frac{1}{2}mv^2 = (60\text{ kg})g(50\text{ m}) - \frac{1}{2}(60\text{ kg})(10\text{ m/s})^2$
   $= (2.6 \times 10^4\text{ J})(1\text{ kcal}/4186\text{ J}) = \underline{6.2\text{ kcal}}$
8. $E_k/E = \frac{1}{2}mv^2/mgh = v^2/2gh = (6.6\text{ m/s})^2/2g(2.5\text{ m}) = 0.89$ or 89%
   Percent lost = 100% − 89% = <u>11%</u>
9. $(100\text{ kcal/h})(4186\text{ J/kcal})(1\text{ h}/3600\text{ s}) = \underline{116\text{ J/s (W)}}$
10. $1200\text{ W} = (1200\text{ J/s})(1\text{ kcal}/4186\text{ J}) = \underline{0.287\text{ kcal/s}}$

11. (100 kcal) (3.968 Btu/kcal) = 397 Btu

12. (12,000 Btu) (0.25 kcal/Btu)(2.0 h) = 6.0×10^3 Btu

13. $H = mc\,\Delta T$ = (0.50 kg)(1.0 kcal/kg-C°)(10 C°) = 5.0 kcal

14. $H = mc\,\Delta T$ = (1.0 kg)(4186 J/kg-C°)(100 C°) = 4.2×10^5 J

15. (a) $H = mc\,\Delta T$ = (1.0 kg)c(80 C°) = 80 kcal
    (b) (80 kcal) (0.00116 kWh/kcal)(12¢/kWh) = 1.1¢

16. $\Delta T_{Cu}/\Delta T_{Al} = c_{Al}/c_{Cu}$ = 0.22/0.092 = 2.4 greater, copper

17. $m_i = m_w c_w \,\Delta T/L_i$ = (1.6 kg)(1.0 kcal/kg-C°)(100 C°)/(80 kcal/kg) = 2.0 kg

18. m_i = (1.0 kg)c(20 C°)/80 kcal/kg = 0.25 kg

19. $H_1 = mc\,\Delta T$ = (0.500 kg)(0.50 kcal/kg-C°)(10 C°) = 2.5 kcal
    $H_2 = mL_f$ = (0.500 kg)(80 kcal/kg) = 40.0 kcal
    H_3 = (0.500)(1.00 kcal/kg-C°)(20 C°) = 10.0 kcal
    Total = 52.5 kcal

20. H_1 = (0.200 kg)(0.50 kcal/kg-C°)(10 C°) = 1.0 kcal
    H_2 = (0.200 kg)(540 kcal/kg) = 108 kcal
    H_3 = (0.200 kg)c(100 C°) = 20 kcal
    H_4 = (0.200 kg)(80 kcal/kg) = 16 kcal
    Total = 145 kcal

21. $\varepsilon_{th} = (H_h - H_c)/H_h$ = (300 kcal – 180 kcal)/(300 kcal) = 0.40 or 40%

22. $W = \varepsilon_{th} H_h$ = (0.25)(100 kcal) = (25 kcal) (4186 J/kcal) = 1.0×10^5 J

23. Less than $\varepsilon_c = 1 - (T_c/T_h)$ = 1 – (373 K/543 K) = 0.31 or 31%

24. ε_c = 1 – (373 K/623 K) = 0.40 or 40%
    Efficiency of 42% not practical.

ANSWERS TO RELEVANCE QUESTIONS

5.1 Temperature range because of possible injury (frost bite and burns).

5.2 Generally, we relate "heat" to high temperatures. With a period of high temperatures [much sunlight (energy) received], the earth and atmosphere warm to where we are uncomfortable in the summer "heat"—unless, of course, we have air conditioning to remove heat energy from the home so that we are comfortably "cool." Humidity also plays a role (see Chapter 25 and 5.3 below).

5.3 The evaporation of perspiration, a phase change requiring latent heat, removes heat energy from our bodies and is a cooling process. Here again, humidity plays an important role (Chapter 25). On a humid day, the water vapor in the air prevents evaporative cooling, and we say it is "hot and humid."

5.4 No. The work done to remove the heat for the ice to form creates more entropy elsewhere, so that there is a net increase in entropy.

5.5 No. The chief mechanism of heat removal by a radiator is conduction and convection as a result of coolant circulation in the engine cooling system and forced air blown through the radiator.

ANSWERS TO STUDY GUIDE QUIZ

Multiple-Choice Questions

1. b	2. c	3. c	4. d	5. b
6. d	7. a	8. a	9. a	10. b

Short-Answer Questions

1. Temperature is a measure of the hotness or coldness of a substance, but more specifically it is a measure of the average kinetic energy of the molecules in a substance.
   Heat is a form of energy and is defined as the total internal energy, both kinetic and potential, of the molecules contained in a sample of material.
2. Matter can be found in three distinct forms—solids, liquids, and gases—which are referred to as the three phases of matter. When the molecules of a substance take on a different bonding structure that changes them from one of these forms into another, we say that a phase change has occurred. Energy, usually in the form of heat, must be added to or removed from a system for a phase change to occur.
3. The Celsius temperature scale is based on the reference points at which pure water freezes and boils. If a mercury-in-glass thermometer is placed in boiling water and the position of the mercury is marked on the glass, and then the same thermometer is placed in a mixture of ice and water and the position of the mercury is again marked, these two points can be used to calibrate the 100 degrees Celsius and the zero degrees Celsius points on the thermometer. If the distance between these two points is then divided into 100 equal parts, the thermometer will be calibrated for use on the Celsius temperature scale.
4. Heat is transferred through solids primarily from one molecule to the next in a process called *conduction.*
5. 5/9 (72°F – 32°) = 22°C   Then: 22°C + 273° = 295 K
6. $Q = mc\ \Delta T = 0.30 \text{ kg} \times 4186 \text{ J/kg-C°} \times (87\text{°C} - 22\text{°C}) = 1256 \times 65 \text{ J} = 82{,}000 \text{ J}$
7. Latent heat of vaporization is the heat needed to change one kilogram of a particular substance from the liquid phase to the gas (vapor) phase or from gas back to liquid at the boiling point. Heat must be added to change a liquid into a gas, but heat is released when the transformation is in the other direction.
8. $Q = mL_f = 12.0 \text{ kg} \times 80 \text{ kcal/kg} = 960 \text{ kcal}$
9. The inside and the outside of the inner container of a Thermos bottle are silvered so that heat energy in the form of infrared radiation will be reflected from these surfaces and heat will not be transferred into or out of the inner container by the process of radiation.
10. The freezing point of water decreases slightly as the pressure increases, but the boiling point of water increases with increasing pressure.

Chapter 6

Waves

The concepts of waves, sound, and light are discussed in this chapter. Because most information about our environment comes to us by means of waves (see Section 1.1, "The Senses"), the general properties of waves are studied to prepare the student for the many physical concepts that involve waves.

Emphasize Section 6.2 on electromagnetic waves. It is by means of electromagnetic waves that information in the study of atomic structure (Chapter 9), as well as astronomical data, are obtained. Later, in Chapter 9, the dual nature of light is presented, and an understanding of the wave nature of light is important. Also, the Doppler effect should be explained in detail because of its important role in the theory of the expanding universe.

DEMONSTRATIONS

There are many demonstrations that illustrate waves, sound, and the Doppler effect. For the concept of transverse waves, a long length of rubber hose is very useful. The demonstration of longitudinal waves can be made with a toy Slinky.

A variety of demonstrations are available for sound and electromagnetic waves. Consult the Teaching Aids section at the back of this *Guide* for sources of these. Sound demonstrations are particularly easy and interesting. Illustrate the production of sound by several methods. Perform the ringing-bell demonstration in a bell jar from which the air can be evacuated. The Doppler effect can be demonstrated using an ordinary hair dryer. Use an oscilloscope for graphic representation of different sounds, including speech.

Standing waves can be demonstrated using a mechanical vibrator and a piece of string with a suspended weight. Also, a Kundt's tube may be used to demonstrate standing-wave nodes and antinodes.

ANSWERS TO REVIEW QUESTIONS

1. c
2. a
3. Longitudinal wave: particle displacement parallel to wave velocity—sound wave in air. Transverse wave: particle displacement perpendicular to wave velocity—light wave.
4. (a) m.
   (b) Hz (or 1/s).
   (c) s.
5. The energy input.

6. b
7. b
8. Radio waves, microwaves, infrared, visible, ultraviolet, X-rays, and gamma rays.
9. Longer wavelength: red end. Higher frequency: blue end.
10. No, they are electromagnetic waves.
11. Light: $\lambda = 400$ nm–700 nm; sound: $\lambda \approx 10^{-1}$ m to 1 m. (Calculate or from table.)
12. d
13. a
14. (a) Frequency.
    (b) Intensity or amplitude.
    (c) Harmonics or overtones.
15. No, it is above the audible range. Ultrasonic cleaning, welding, echograms, etc.
16. Interference distorts quality.
17. No. The dB scale is not linear. (An increase of 3 dB doubles intensity.)
18. a
19. b
20. (a) Shortened.
    (b) Lengthened.
21. (a) Shift to a higher frequency. (b) Shift to a lower frequency.
22. The crack of a whip is caused by the end of the whip exceeding the speed of sound (a mini-sonic boom).
23. Detection involves the reflection of waves from an object. Speed and ranging involve the Doppler shift of the reflected waves.
24. c
25. d
26. There must be a node at each end.
27. String: any number n. Pendulum: one.
28. Maximum energy transfer to the system when driven at a resonance frequency.
29. The linear mass density and tension applied to a string. By varying the length (with a finger on the string) and/or tension (tuning).

ANSWERS TO CRITICAL THINKING QUESTIONS

1. It depends on your definition of sound. Sound as being perceived (heard) waves, no. Sound as being a wave in air, yes.
2. No. There is no atmosphere on the Moon. Communication is by radio waves, which are electromagnetic waves that require no medium for propagation.
3. (a) No effect, no shift in waves because of zero relative velocity. Source and observer are relatively stationary. (b) Sonic boom.
4. Yes. Air in the cabin is traveling with the plane and is relatively stationary.
5. Because of standing waves (overtones) set up between the shower walls.

ANSWERS TO EXERCISES

1. $f = 1/T = 1/(0.25\ \text{s}) = \underline{4.0\ \text{Hz}}$

2. $T = 1/f = 1/(10 \times 10^3\ \text{Hz}) = \underline{0.00010\ \text{s}}$

3. (a) $f = v/\lambda = (2.0\ \text{m/s})/(1.5\ \text{m}) = \underline{1.3\ \text{m}}$
   (b) $T = 1/f = 1/(1.3\ \text{Hz}) = \underline{0.769\ \text{s}}$

4. $\lambda = c/f = (344\ \text{m/s})/(2000\ \text{Hz}) = \underline{0.172\ \text{m}}$

5. $\lambda = c/f = (3.00 \times 10^8\ \text{m/s})/(6.50 \times 10^5\ \text{Hz}) = \underline{4.62 \times 10^2\ \text{m}}$

6. $\lambda = (3.0 \times 10^8\ \text{m/s})/(10^{18}\ \text{Hz}) = \underline{3.0 \times 10^{-10}\ \text{m, or } 0.30\ \text{nm}}$

7. $f = c/(= (3.00 \times 10^8\ \text{m/s})/(420 \times 10^{-9}\ \text{m}) = \underline{7.14 \times 10^{14}\ \text{Hz}}$

8. $f_l/f_s = (7.5 \times 10^{14}\ \text{Hz})/(2.0 \times 10^4\ \text{Hz}) \approx \underline{10^{10}}$

9. $\lambda = v/f = (344\ \text{m/s})/(50 \times 10^3\ \text{Hz}) = \underline{6.9 \times 10^{-3}\ \text{m}}$

10. $\lambda_1 = v/f = (344\ \text{m/s})/(20\ \text{Hz}) = \underline{17\ \text{m}}$
    $\lambda_2 = v/(2.0 \times 10^4\ \text{Hz}) = \underline{0.017\ \text{m}}$

11. $d = vt = (1/3\ \text{km/s})(4.0\ \text{s}) = \underline{1.3\ \text{km}}$
    $d = (1/5\ \text{mi/s})(4.0\ \text{s}) = \underline{0.80\ \text{mi}}$

12. $d = vt = (1/3\ \text{km/s})(9.0\ \text{s}) = 3.0\ \text{km}$
    $t = d/v_s = (3.0\ \text{km})/(12\ \text{km/h}) = 1/4\ \text{h} = \underline{15\ \text{min}}$

13. $\Delta = 20\ \text{dB} = 2\ \text{B}$, and $10^2 = \underline{100}$

14. $10{,}000 = 10^4$, and 4 B or 40 dB increase to $\underline{120\ \text{dB}}$

15. $1/10 = 10^{-1}$, and –1 B or –10 dB decrease to $\underline{108\ \text{dB}}$

16. (a) 3 dB decrease to $\underline{120\ \text{dB}}$
    (b) 10^{-2}, and –2 B or –20 dB decrease to $\underline{103\ \text{dB}}$

17. $f_1 = v/2L = (240\ \text{m/s})/[2(1.0\ \text{m})] = \underline{120\ \text{Hz}}$

18. $f_1 = 120\ \text{Hz}$, $f_2 = 2f_1 = \underline{240\ \text{Hz}}$

19. $f_2 = 2f_1 = 2(256\ \text{Hz}) = \underline{512\ \text{Hz}}$
    $f_3 = 3f_1 = \underline{768\ \text{Hz}}$

20. $3f_1 = 660\ \text{Hz}$, and $f_1 = 220\ \text{Hz}$, then $f_4 = 4f_1 = \underline{880\ \text{Hz}}$

21. $\lambda_n = 2L/n$, $n = 1, 2, 3, \ldots$
    $f_n = v/\lambda_n = nv/2L$

22. $f_1 = v/2L$ and $L = v/2f_1 = (344\ \text{m/s})/[2(440\ \text{Hz})] = \underline{0.391\ \text{m}}$

23. $\lambda_m = 4L/m$, $m = 1, 3, 5, \ldots$
    $f_m = v/\lambda_m = mv/4L$

24. (a) $f_1 = v/4L = (344\text{ m/s})/[4(2.0\text{ m})] = \underline{43\text{ Hz}}$
    (b) $f_5 = 5f_1 = 5(43\text{ Hz}) = \underline{215\text{ Hz}}$

ANSWERS TO RELEVANCE QUESTIONS

6.1 Yes, the sense of touch can be used to detect waves (vibrations) in objects and fluids (e.g., water waves).

6.2 Through sunburns and suntans. UV can also cause blindness (not a good detection method).

6.5 The cavities in the throat and nasal regions. The combinations of resonances in these cavities give different people different voice sounds.

ANSWERS TO STUDY GUIDE QUIZ

Multiple-Choice Questions

1. c	2. b	3. d	4. d	5. a
6. c	7. a	8. d	9. a	10. d

Short-Answer Questions

1. Any four of the following: radio waves, microwaves, infrared waves (IR), ultraviolet (UV), X-rays, or gamma rays.

2. Wave speed is the rate at which wave energy travels through any given material or, in the case of electromagnetic waves, through vacuum. Wave speed can be determined by measuring the distance that the wave travels and dividing this distance by the time it takes for the wave to travel that far.

   Wavelength is the distance from any point on a wave to the adjacent point with similar oscillation. An example would be from one crest of a wave to the next crest. This is the distance of one complete "wave," shown in Figure 6.5 in the textbook as λ.

3. Sound propagates as a series of compressions and rarefactions in the air as the air molecules themselves are pushed closer together than usual and then are pulled farther apart. The resulting wave motion transfers sound energy from a sound source throughout the surrounding area. When these wave oscillations reach our ears, we "hear" the sound that is being produced.

4. A standing wave is a fixed pattern of nodes and antinodes that are produced when an initial wave interacts with its own reflected waveform that is returning from a reflection at some boundary, such as the stationary end of a stretched string. This interaction can be constructive, when the wave and its reflected portion are in phase, at which point antinodes are produced. If the initial wave and its reflected portion are out of phase, destructive interference occurs and a node is produced. Since the fixed pattern does not change with time, it is referred to as a standing wave.

5. The frequency of the sound wave produced by the siren is decreased as the vehicle travels away from an observer. This results in a lower pitch being heard when the vehicle is moving away from the observer. This is known as the Doppler effect.

6. The upper limit of frequency that can be detected by the average human ear is 20,000 Hz, whereas the lowest frequency that can normally be heard is 20 Hz.

7. Since period is simply the reciprocal of the frequency,

$$T = 1/f = 1/(880\text{ Hz}) = 0.00114\text{ s}$$

8. Frequency of the 2nd harmonic = 2 × fundamental frequency = 2(880 Hz) = 1760 Hz
Frequency of the 3rd harmonic = 3 × fundamental frequency = 3(880 Hz) = 2640 Hz

9. Since the speed of light in vacuum = 3.00×10^8 m/s,

$$\lambda = v/f = (3.00 \times 10^8 \text{ m/s})/(3.50 \times 10^{14} \text{ Hz}) = 8.57 \times 10^{-7} \text{ m}$$

10. The length of the string is 1/2 of the fundamental wavelength (see Figure 6.17a in the textbook), and so

$$\lambda = 2(0.750 \text{ m}) = 1.50 \text{ m}$$

$$f = v/\lambda = (792 \text{ m/s})/(1.50 \text{ m}) = 528 \text{ Hz}$$

This frequency is that of the fifth C found on the piano (see Figure 6.19 in the textbook).

Chapter 7

Wave Effects and Optics

The six major properties of waves (reflection, refraction, dispersion, diffraction, interference, and polarization) are presented and discussed in this chapter. Introduce the properties to the students by means of demonstrations. Reflection can be illustrated with a plane mirror. Also, note that everything in the classroom is seen by reflection except the light source. A pencil or a small ruler in a glass of water is a good demonstration of refraction, and a prism can be used to illustrate dispersion.

Viewing a candle through the slit between two fingers is good for illustrating diffraction, as is viewing a candle flame through a feather. Interference of light waves is best illustrated by Young's experiment. If an experiment with a spectroscope is not done in the laboratory, the instrument should be presented and demonstrated in class. Polarized light is easily demonstrated with crossed Polaroids. (Polarizing sunglasses may be used.)

Considerable class time should be spent in constructing ray diagrams and locating real and virtual images for different types of spherical mirrors and lenses.

DEMONSTRATIONS

There are many demonstrations for illustrating waves, sound, and the basic laws of optics. A few have been mentioned in the introduction above. The ripple tank is an excellent piece of apparatus for the production and projection of water waves. The apparatus can be used to study reflection, refraction, diffraction, and interference of waves.

Distribute replica diffraction gratings to the students in class and let them see the spectrum from an incandescent light source. Demonstrate polarization with linear polarizers and double refraction with an Iceland spar crystal. If available, a laser is a spectacular demonstration in itself, and many optical properties can be demonstrated with commercial laser kits.

ANSWERS TO REVIEW QUESTIONS

1. d
2. d
3. Diffuse reflection is from a rough surface, and the rays are not reflected parallel, although the law of reflection applies to each ray and surface. The reflection from this page is a good example.
4. At the same rate as you walk toward the mirror.
5. c

6. c
7. The bending of waves around the corners of objects. For light, there is a bending or change in direction on entering a medium because of a change in the wave speed.
8. The frequency remains the same, but marchers entering a muddy region or marking time do not take as many steps and slow up. This causes a change in the direction of the column.
9. (a) Toward the normal.
   (b) Away from the normal.
10. Because nothing can exceed the speed of light c, and $n = c/_{\mathrm{m}}$.
11. Dispersion is the separation of light into its components, e.g., by a prism. It allows for spectral identification of substances.
12. Brilliance is due to internal reflection, and "fire" is due to dispersion.
13. a
14. c
15. c
16. For $\lambda \leq 0.50$ cm.
17. The wavelengths of sound satisfy the diffraction condition. Light waves have very short wavelengths.
18. (a) Amplitudes will add.
    (b) Amplitudes cancel, but not completely.
19. Starting with uncrossed polarizing sheets, a darkening, a lightening, a darkening, and a lightening.
20. a
21. b
22. The distance between the vertex and the center of curvature is the radius of curvature R, and half this distance is the focal length, which defines the focal point at $R/2$.
23. Real images are formed on the object side of the mirror and can be displayed or focused on a screen. Virtual images are formed "behind" or "inside" the mirror, and cannot be formed on a screen.
24. (a) Concave: real, $D_o > f$.
    (b) Convex: always virtual.
25. a
26. c
27. (a) Convex: real, $D_o > f$; virtual, $D_o < f$.
    (b) Concave: always virtual.
28. Because the lens or lens system gives an inverted image. Focusing involves adjusting the object distance so that a sharp image is formed on the screen.
29. The image of an object inside the focal point forms a magnified, virtual image, which can act as an object for the eye.

ANSWERS TO CRITICAL THINKING QUESTIONS

1. Like a spherical, convex mirror.
2. Yes. Study Fig. 7.8 in reverse. The marchers reaching the solid ground first would be able to take bigger steps, which turn the column away from the normal.

3. The fish would see the 360° above-water panorama in a circular cone defined by the critical angle.
4. (a) Yes, it reflects TV waves to its focal point where the receiver is placed.
   (b) The wavelengths are sufficiently long that the mesh surface acts as a mirror.
5. The convex mirror produces a reduced image, and the smaller image may be interpreted as being more distant than it actually is.

ANSWERS TO EXERCISES

1. $\theta_i = \theta_r = \underline{30°}$
2. $\theta_r = 90° - 30° = \underline{60°}$
3. No. Angles bisect to give half height for any distance.
4. 76 in./2 = 38 in., and 62 in./2 = 31 in. So, $\Delta = \underline{7 \text{ in.}}$
5. $c_m = c/n = (3.00 \times 10^8 \text{ m/s})/2.42 = \underline{1.24 \times 10^8 \text{ m/s}}$
6. $n = c/c_m = (3.00 \times 10^8 \text{ m/s})/(1.60 \times 10^8 \text{ m/s}) = \underline{1.88}$
7. $c_m/c = 1/n = 1/1.52 = 0.658\ (\times 100\%) = \underline{65.8\%}$
8. $c_m/c = 0.413 = 1/n$, and $n = 1/0.413 = \underline{2.42\text{, diamond}}$
9. Sketch ray diagram.
10. Sketch ray diagrams. Reduced image becomes larger as objects move toward the mirror approaching the center of curvature, with $M = 1$ at that point. Continues to become larger as object approaches focal point, and image becomes virtual inside focal point.
11. $D_i = D_o f/(D_o - f) = (20 \text{ cm})(60 \text{ cm})/[(20 - 60) \text{ cm}] = -30 \text{ cm}$
    $M = -D_i/D_o = -(-30 \text{ cm})/(20 \text{ cm}) = \underline{1.5}$
12. (a) $D_i = (20 \text{ cm})(15 \text{ cm})/(20 - 15) \text{ cm}] = \underline{60 \text{ cm}}$
    (b) $M = (-60 \text{ cm})/(20 \text{ cm}) = \underline{-3.0}$
    $h_i = 3.0h_o = (3.0)(3.0 \text{ cm}) = \underline{9.0 \text{ cm}}$
13. $D_i = D_o f/(D_o - f) = 15 \text{ cm}(-10 \text{ cm})/[15 \text{ cm} - (-10 \text{ cm})] = \underline{-6.0 \text{ cm (virtual)}}$
    $M = -D_i/D_o = -(-6.0 \text{ cm})/(15 \text{ cm}) = \underline{0.40 \text{ (upright)}}$
14. $D_i = (20 \text{ cm})(-4.0 \text{ cm})/[20 \text{ cm} - (-4.0 \text{ cm})] = \underline{-3.3 \text{ cm (virtual)}}$
    $M = -(-3.3 \text{ cm})/20 \text{ cm} = \underline{0.17 \text{ (upright)}}$
15. Sketch ray diagram.
16. Sketch ray diagrams. Reduced image becomes larger as object moves toward lens approaching the $2f$ position, with $M = 1$ at that point. Then $M > 1$ as object approaches the focal point, and image becomes virtual inside the focal point.
17. $D_i = D_o f/(D_o - f) = (45 \text{ cm})(20 \text{ cm})/[(45 - 20) \text{ cm}] = \underline{36 \text{ cm}}$
    $M = -D_i/D_o = (-36 \text{ cm})/(45 \text{ cm}) = \underline{-0.80 \text{ (inverted)}}$
    $h_i = (0.80)h_o = (0.80)(5.0 \text{ cm}) = \underline{4.0 \text{ cm}}$

18. (a) $D_i = (8.2\text{ cm})(8.0\text{ cm})/[(8.2 - 8.0)\text{ cm}] = \underline{328\text{ cm}}$
    (b) $M = (-328\text{ cm})/(8.2\text{ cm}) = -40$, and $\underline{40\text{ in.} \times 40\text{ in.}}$

19. $D_i = D_o f/(D_o - f) = (15\text{ cm})(15\text{ cm})/[(15 - 15)\text{ cm}] = \underline{\infty}$
    $M = -D_i/D_o = -\infty/15 = \underline{\infty}$

20. $D_i = (6.0\text{ cm})(8.0\text{ cm})/[(6.0 - 8.0)\text{ cm}] = -24\text{ cm}$
    $M = -(-24\text{ cm})/(6.0\text{ cm}) = \underline{4.0}$

21. $D_i = D_o f/(D_o - f) = (10\text{ cm}) -(-20\text{ cm})/[10\text{ cm} -(-20\text{ cm})] = \underline{-6.7\text{ cm (virtual)}}$
    $M = -D_i/D_o = -(-6.7\text{ cm})/(10\text{ cm}) = \underline{0.67\text{ (upright)}}$

22. No. $D_o = 25\text{ cm}$ $D_i = (25\text{ cm})(-20\text{ cm})/[25\text{ cm} - (-20\text{ cm})] = -11\text{ cm}$
    $M = -(-11\text{ cm})/(25\text{ cm}) = \underline{0.44}$
    $D_o = 15\text{ cm}$ $D_i = (15\text{ cm})(-20\text{ cm})/[15\text{ cm} - (-20\text{ cm})] = -8.6\text{ cm}$
    $M = -(-8.6\text{ cm})/(15\text{ cm}) = \underline{0.57}$

23. 2.20 cm focal length gives near point. 2.40 cm at infinity.
    $D_o = D_i f/(D_i - f) = (2.40\text{ cm})(2.20\text{ cm})/[(2.40 - 2.20\text{ cm})] = \underline{26.4\text{ cm}}$

24. $1/f_1 = 1/D_o + 1/D_i = 1/(25\text{ cm}) + 1/(2.5\text{ cm}) = 0.44$, and $f_1 = 2.3\text{ cm}$
    $1/f_2 = 1/\infty + 1/(2.5\text{ cm}) = 0.44$, and $f_2 = \underline{2.5\text{ cm}}$

ANSWERS TO RELEVANCE QUESTIONS

7.1 With the dark background, the reflected light can be seen and the window acts as a mirror. During the day, the light coming through the window masks the reflected light.

7.2 We tend to think of light as traveling in a straight line, but the light coming from the fish to your eye is refracted or bent (away from the normal) at the water-air interface. Hence the fish is not at the extended straight-line position of the light ray coming to your eye. (Make a sketch and show this.)

7.3 Use the first pair as an analyzer and rotate one of the lenses of this pair while looking through it and one of the lenses of the other pair of glasses. If the second pair is polarizing, you will see crossed Polaroid darkening.

7.4 The front of the spoon acts as a concave mirror, and the back surface as a convex mirror. Outside the focal point of the front side, you see an inverted image; and you see an upright image for the back side (which is always produced for a convex mirror). You can see an upright image in the front side if you bring the spoon close enough so that you as the object are inside the focal point. Try it and see.

ANSWERS TO STUDY GUIDE QUIZ

Multiple-Choice Questions

1. c 2. a 3. c 4. c 5. d
6. a 7. a 8. c 9. d 10. c

Short-Answer Questions

1. Regular reflection occurs from a smooth surface like a mirror, and the result is the formation of a clear image of the source of any light striking the surface. Light striking a rough surface like a brick wall will still be reflected (diffuse reflection), but the rays will be dispersed and will now show a clear image of the light source.
2. The law of reflection says that for any light ray striking a reflecting surface at a given angle, as measured from the perpendicular (normal) line drawn to the surface, the angle of reflection for the ray, as measured from the normal line, will be exactly the same. In other words, the angle of incidence for a light ray is always equal to its angle of reflection. This is true for both rough and smooth surfaces; the diffuse effect on rough surfaces is caused because the normal lines for different rays are not in the same direction, even though they are all perpendicular to the surface at the individual points of contact.
3. The index of refraction of a medium is the ratio of the speed of light in vacuum to the speed of light traveling through that medium.

   Focal length is the distance between a lens or mirror and a point at which an image of an object that is a long way (infinite distance) from the lens or mirror can be formed. In actual practice, this "long way" must be on the order of 20 or more times the focal length for the image to form at the exact focal point of the lens or mirror.

   Polarization is the removal of variations in the direction in which the electric and magnetic fields oscillate in an electromagnetic wave so that all electric and magnetic field variations are in one orientation; see Figure 7.21 in the textbook. This is actually a preferential orientation of the field vectors and can be either partial or complete. If the field vectors are all in the same plane, the light is said to be linearly polarized.
4. White light from the sun passes through raindrops that refract and reflect the light in such a way that the white light is separated into its component colors. When the light emerging from many raindrops is observed from a certain location on the ground, a bright band of spectral colors appears as an arc, or sometimes only as a partial arc, across the sky. More details on this process are given in the first chapter highlight in the textbook.
5. A light pipe is a solid glass or plastic rod through which light can be passed in such a way that no light is refracted out of the rod through the side walls. This occurs because of total internal reflection of the light rays from the side walls. Total internal reflection is the result of rays striking the surface of the glass or plastic at an angle greater than the critical angle for that material. The critical angle depends on the relative indexes of refraction of the light pipe itself and the material surrounding it, quite often air.
6. $v = c/n = (3.00 \times 10^8 \text{ m/s})/2.42 = 1.24 \times 10^8 \text{ m/s}$
7. When two light waves having the same wavelength strike a spot on a screen in phase with each other, the amplitudes of the two waves add together to produce constructive interference, and a bright spot is observed on the screen.
8. $1/D_i = 1/f - 1/D_o = 1/(7.0 \text{ cm}) - 1/(18.0 \text{ cm}) = 0.1429 - 0.0556 = 0.0873$

   then $D_i = 1/0.873 = 11.4$ cm

   The image is located 11.4 cm on the other side of the lens from the object's position.
9. $M = -D_i/D_o = (-11.4 \text{ cm})/(18.0 \text{ cm}) = -0.636$

   The image will be smaller by about 40% and will be inverted, as shown by the minus sign.
10. As described in the textbook, thin-film interference can explain the colorful displays seen as (1) thin oil films and (2) soap bubbles. This process also explains the bright spectral colors seen in opal, mother-of-pearl, and peacock feathers.

Chapter 8

Electricity and Magnetism

This is an important chapter because knowledge of the basic concepts of electricity and magnetism is necessary in the study of atomic physics. Also, students should have a basic knowledge of electrical circuits because electricity plays such an important role in our lives.

Coulomb's law and magnetic fields should be stressed so as to provide an understanding of some of the underlying concepts in the study of modern physics and chemistry. The chapter is somewhat long, but the study of electric and magnetic phenomena is well worth the effort.

DEMONSTRATIONS

Lecture demonstrations with an electroscope that can be projected on a large screen are very good and create interest.
Do the demonstration shown in Fig. 8.4.
Demonstrate a Van de Graaff generator.
Demonstrate electromagnetic induction. (Commercial apparatuses are available.)
Demonstrate the electric generator and electric motor. (May be constructed or commercially available.)
Demonstrate the compass and dip needle.
Pass around a cheap compass and a bar magnet.
If possible, bring in a phone receiver and show where the various parts are located.

ANSWERS TO REVIEW QUESTIONS

1. b
2. c
3. c
4. Negatively charged electrons, positively charged protons, and neutral neutrons. The latter two are almost 2000 times more massive than the first.
5. Because the electrical force is many orders of magnitude stronger than the gravitational force.
6. A flow of charge, which is measured in amperes (A) or coulomb/second (C/s).
7. Because of attractive electrical forces arising from charging by induction and polarized molecules.
8. Some materials have free electrons for conduction, and others do not.
9. b

10. d
11. Electric potential energy arises due to work done against an electric force. Voltage is electrical energy (or work) per unit charge, $V = W/q$.
12. $V = IR$, volts, amps, and ohms.
13. (a) $P = IV$ (b) $P = I^2R$
14. This is the energy (power) lost because of resistance.
15. b
16. b
17. Direct current has a voltage with constant polarity and flows in only one direction. Alternating current results from a changing polarity, and the current alternates in direction.
18. So that all will have the same voltage and independent paths so they can be operated independently.
19. (a) Opens a circuit by the melting of a metal strip when current is too large.
    (b) Opens a circuit by magnetic or thermal means when current is too large.
    (c) Dedicated grounding wire to prevent object being at high voltage.
    (d) Use of ground wire as a grounding wire.
20. Voltage and body resistance. Don't become part of a circuit.
21. b
22. d
23. Similar. Likes repel, unlikes attract.
24. (a) One that is easily magnetized. Iron, nickel, and cobalt.
    (b) The ferromagnetic iron becomes an induced magnet. Above the Curie temperature, iron loses its magnetism.
25. (a) The magnetic field of a bar magnet with the north magnetic pole near the geographic south pole.
    (b) Declination is the horizontal angular distance between Earth's magnetic field lines or compass direction (magnetic north) and true north. Maps are based on true north, so it is necessary to know the declination to navigate properly.
26. c
27. b
28. Based on electromagnetic phenomena. See text for description.
29. Based on torque on a current-carrying wire. See text for description.
30. Based on electromagnetic induction. Current set up in a wire moving in a magnetic field. See text for description.
31. Electromagnetic induction. Used to step-up and step-down voltage and current so as to reduce loss in electrical power transmission.
32. Less losses in power transmission.
33. Much of the power would be lost through I^2R losses.

ANSWERS TO CRITICAL THINKING QUESTIONS

1. We do not know. Electric charge is simply an associated property of electrons and protons.
2. There would be strong intermolecular forces and matter would be very stable with few phase changes.
3. With one hand in the pocket, you are not likely to grasp a high voltage with both hands and provide a path across your chest.
4. Two magnets are obtained. Creating magnetic monopoles in this manner is not possible. Magnetism is the result of domain alignment.
5. No, not with steady direct current, since there would be no change in the magnetic field and no magnetic induction. Pulsating dc current would have an effect.

ANSWERS TO EXERCISES

1. $n = q/e = 1.00 \text{ C}/1.60 \times 10^{-19} \text{ C} = \underline{6.25 \times 10^{18}}$
2. $q = ne = (10^6)(1.6 \times 10^{-19} \text{ C}) = \underline{1.6 \times 10^{-13} \text{ C}}$
3. $I = q/t = ne/t = (4.8 \times 10^{18})(1.6 \times 10^{-19} \text{ C})/0.25 \text{ s} = \underline{3.1 \text{ A}}$
4. $q = It = (1.50 \text{ A})(6.0 \text{ s}) = \underline{9.0 \text{ C}}$
5. $F = kq_1q_2/r^2 = (9.0 \times 10^9 \text{ N–m}^2/\text{C}^2)(0.50 \text{ C})(2.0 \text{ C})/(3.0 \text{ m})^2$
   $= \underline{1.0 \times 10^{-9} \text{ N, mutually repulsive}}$
6. $F = kpe/r^2 = k(1.6 \times 10^{-19} \text{ C})^2(5.0 \times 10^{-11} \text{ m})^2 = \underline{9.2 \times 10^{-8} \text{ N}}$
   $F = Gm_em_p/r^2 = (6.67 \times 10^{-11})(9.11 \times 10^{-31})(1.67 \times 10^{-27})/(5.0 \times 10^{-11})^2$
   $= \underline{4.1 \times 10^{-47} \text{ N}}$
7. $U = W = 30 \text{ J}$
8. $V = W/q = 30 \text{ J}/0.25 \text{ C} = \underline{120 \text{ V}}$
9. $I = V/R = 120 \text{ V}/40\ \Omega = \underline{3.0 \text{ A}}$
10. $V = IR = (0.50 \text{ A})(24\ \Omega) = \underline{12\ \Omega}$
11. (a) $P = IV = (0.25 \text{ A})(12 \text{ V}) = \underline{3.0 \text{ W}}$
    (b) $R = V/I = 12 \text{ V}/0.25 \text{ A} = \underline{48\ \Omega}$
12. (a) $I = P/V = 0.50 \text{ W}/3.0 \text{ V} = \underline{0.17 \text{ A}}$
    (b) $R = V/I = 3.0 \text{ V}/0.17 \text{ A} = \underline{18\ \Omega}$
13. $E = Pt = (1.5 \text{ kW})(0.50 \text{ h/day})(30 \text{ day}) = 22.5 \text{ kWh} (\times \$0.08/\text{kWh}) = \underline{\$1.80}$
14. $E = (1.0 \text{ kW})(3.0 \text{ h/day})(30 \text{ day}) = 90 \text{ kWh} (\times \$0.10/\text{kWh}) = \underline{\$9.00}$
15. (a) $R = V/I = 110 \text{ V}/10 \text{ A} = \underline{11\ \Omega}$
    (b) $P = IV = (10 \text{ A})(110 \text{ V}) = \underline{1100 \text{ W}}$
16. (a) $I = P/V = 100 \text{ W}/120 \text{ V} = \underline{0.83 \text{ A}}$
    (b) $R = V/I = 120 \text{ V}/0.83 \text{ A} = \underline{144\ \Omega}$
    (c) $P = E/t = 100 \text{ J/s}$, or 100 J each second.

17. (a) $I = V/R = 12\ V/90\ \Omega = \underline{0.13\ A}$
    (b) $P = IV = (0.13\ A)(12\ V) = \underline{1.6\ W}$
18. $1/R_p = 1/20\ \Omega + 1/30\ \Omega + 1/40\ \Omega$, and $R_p = 9.2\ \Omega$
    (a) $I = V/R = 12\ V/9.2\ \Omega = \underline{1.3\ A}$
    (b) $P = IV = (1.3\ A)(12\ V) = \underline{156\ W}$
19. (a) $1/R_p = 1/10\ \Omega + 1/15\ \Omega$, and $R_p = \underline{6.0\ \Omega}$
    (b) $R = 8 \times 6.0\ \Omega = 48\ \Omega$, and $I = V/R = 120\ V/48\ \Omega = \underline{2.5\ A}$
20. (a) $R_s = R_1 + R_2 + R_3 = 60\ \Omega + 30\ \Omega + 20\ \Omega = \underline{110\ \Omega}$
    (b) $R_p = R_1R_2/(R_1 + R_2) = (110\ \Omega)(50\ \Omega)/(110\ \Omega + 50\ \Omega) = 34\ \Omega$
    $I = V/R = 12\ V/34\ \Omega = \underline{0.35\ A}$
21. (a) $V_2 = (N_2/N_1)V_1 = (300/50)(12\ V) = \underline{72\ V}$
    (b) $I_2 = (N_1/N_2)V_1 = (50/300)(3.0\ A) = \underline{0.50\ A}$
22. (a) $N_1 > N_2$, step–down
    (b) $V_2 = (50/250)(100\ V) = \underline{20\ V}$
    $I_2 = (250/50)(0.25\ A) = \underline{1.25\ A}$
23. $N_2 = (V_2/V_1)N_1 = (220\ V/4400\ V)(1000) = \underline{50\ turns}$
24. $N_2 = (40{,}000\ V/240{,}000\ V)(900) = 150$ turns

ANSWERS TO RELEVANCE QUESTIONS

8.1 Electrostatic charging of clothes causes the clothes to cling together and attract foreign objects (try cat hair). A fine spray of water or some commercial agent on the clothes allows the static charge to be conducted away.

8.2 Low resistance. Although electrical resistance generates heat, the lack of resistance allows more current to flow and the I^2R losses are greater because of the squaring effect of the current.

ANSWERS TO STUDY GUIDE QUIZ

Multiple-Choice Questions

1. a	2. b	3. c	4. b	5. d
6. a	7. b	8. a	9. d	10. c

Short-Answer Questions

1. Electrostatic charges of the same sign will repel each other, and so the two negatively charged rods will try to push each other apart.
2. Electrostatic interactions are described by the *law of charges*, which states that unlike charges attract, but like charges repel. This means that two negatively charged objects or two positively charged objects will repel each other, whereas a positively charged object will attract a negatively charged object.
3. When a rubber balloon is rubbed with wool, some of the electrons from the rubber atoms are removed, and the balloon is left with a net positive charge. When this charged object gets close to a neutral surface, such as a wall or ceiling, an opposite electrostatic charge is sometimes induced on the surface, resulting in a tendency to attract the balloon.

4. Electric current is the movement of electric charge. This movement can consist of either positive or negative charge transfer through liquids or gases, but is generally the transfer of negative charge through solids in the form of electrons that can move in conductors such as metals. Electric current through a wire as the amount of electric charge (in coulombs) that moves through the wire per second. The unit for electric current is the ampere.

5. Ohm's law states that $V = IR$, where V is the potential difference or voltage, R is the electrical resistance, and I is the current flowing through the resistance.

6. Each light bulb in your home is connected in parallel with the other light bulbs so that the same voltage (110 VAC) is applied across each one, and each light bulb can draw as much current as it needs according to Ohm's law, as determined by its individual internal resistance. A fuse and switch are usually connected in series with the bulbs to prevent excess current from flowing if there is a short in the circuit and to allow the bulb to be easily turned on and off as needed.

7. In a dc circuit, the electric current always flows through the circuit in the same direction—from the positive terminal of a battery, for example, toward the negative terminal. The current in an ac circuit, on the other hand, flows first in one direction and then in the other in a repeatable manner. Our household ac circuits operate at a rate of 60 cycles per second.

8. A three-prong grounded plug and socket system provides an additional separate ground wire that protects the user of electrical equipment by providing a backup safety path so that electric current can return to ground without endangering the user if a short or other electrical malfunction occurs in the wiring. Two of the connections are the same as those in a standard two-prong plug, but the third connection has a separate wire that goes all the way back to solid ground to provide an alternative path for the current if the primary ground wire in the plug is cut or damaged.

9. A transformer is made up of two coils of wire on a ferromagnetic core that can be used to raise or lower the electrical voltage supplied to a circuit. If there are more windings on the secondary of the transformer than on the primary, the transformer will increase the voltage; if there are fewer windings on the secondary, the voltage will be lower than that supplied to the primary. A transformer can, therefore, be used to supply high voltage for TV picture tubes or low voltage for calculators or tape players from the same primary 110-VAC source. The electrical circuit must be alternating current (ac) for the transformer to work. Transformers cannot be used to adjust the voltage in a dc circuit, such as one whose power is supplied by a battery.

10. A ferromagnetic material has an internal structure that allows it to form domains in which all of the electron spins of the constituent atoms are in the same direction. Ferromagnetic materials also have more unpaired electrons per atom that can contribute to the overall magnetic effect produced by this alignment. When all of the electron spins are oriented in the same direction, the material is saturated and the strongest magnetic effects that are possible are produced.

Chapter 9

Atomic Physics

The subject matter of atomic physics includes some of the revolutionary concepts and developments of twentieth-century physics. First, for background, we review the models of the atom up to 1911. The dual nature of light is then introduced, followed by the Bohr theory of the hydrogen atom. Quantum applications—microwave ovens, X-rays, lasers—are discussed, followed by the concept of the dual nature of matter. The quantum mechanical model of the atom is treated, including the electron configurations for the first 20 elements. The chapter ends with a brief discussion of Heisenberg's uncertainty principle. The chapter Highlight discusses the phenomena of fluorescence and phosphorescence.

We have kept the discussions of these difficult topics as simple and straightforward as possible. The mathematical treatment has been simplified and reduced from that in earlier editions. Sections 9.1, 9.2, and 9.3 are basic material. Instructors should use their judgment as to what additional parts of the chapter to cover and emphasize.

DEMONSTRATIONS

Many of the topics in this chapter are difficult to demonstrate. Fluorescence and phosphorescence may be demonstrated with an ultraviolet lamp and appropriate samples from supply houses. Students especially seem to like finding that fluorescent chalk has been used to write their homework assignment on the board.

A common laser may be demonstrated, but use appropriate safety precautions. A food item may be heated in a microwave oven to demonstrate that the center remains cold if sufficient time for conduction of heat to the center is not allowed. For additional ideas on demonstrations, refer to the publications listed in the appendix of this *Guide*.

Appropriate films, videos, and transparencies should be especially helpful in this chapter. Refer to the catalogs of the audio-visual suppliers listed in the appendix of this *Guide*.

ANSWERS TO REVIEW QUESTIONS

1. c
2. a
3. Dalton's model of the atom as a featureless "billiard ball" was abandoned for Thomson's "plum pudding" model after electrons were found to be a constituent of all atoms.
4. Thomson found that electrons were deflected by electrical and magnetic fields in such a way that their electrical charge had to be negative.

5. Thomson's "plum pudding" model pictured the atom as consisting of electrons randomly positioned like raisins in an otherwise homogeneous mass of positively charged "pudding." Thomson's model was abandoned when Rutherford discovered that each atom had a tiny core, or nucleus, in which 99.9% of the mass and all of the positive charge were concentrated and around which the electrons circulated.
6. a
7. d
8. The wave nature of light is shown by phenomena such as diffraction, interference, and polarization. The particle nature of light is shown by phenomena such as the photoelectric effect.
9. Something is quantized when it is restricted to certain discrete values rather than having a continous range of values.
10. A *proton* is a postively charged subatomic particle found in the nuclei of atoms. A *photon* is a quantum, or "particle," of electromagnetic radiation.
11. Frequency and wavelength are inversely proportional.
12. A photon of yellow light has less energy, lower frequency, and longer wavelength than a photon of green light.
13. Only light above a certain frequency causes electrons to be ejected, and thus the photon must have a certain minimum energy to cause ejection; $E = hf$—the higher the frequency, the greater the photon energy.
14. c
15. c ($n = 7.5$ is an impossible value.)
16. A radius of 0.053 nm would mean a diameter of 0.106 nm for a hydrogen atom.
17. Bohr postulated that an orbiting electron does not radiate energy when it is in an allowed, discrete orbit, but does so only when it makes a quantum jump, or transition, from one allowed orbit to another.
18. The ground state for an electron is the lowest energy state that it can be in. Energy states above the ground state are called excited states.
19. Four visible lines—red, blue-green, and two violet—make up the line emission spectrum of hydrogen, as shown in Figs. 9.11b and 9.17.
20. The four discrete lines in the emission spectrum of hydrogen correspond to electron transitions down to $n = 2$ from $n = 6, 5, 4$, and 3.
21. The four discrete lines in the absorption spectrum of hydrogen correspond to electron transitions up to $n = 6, 5, 4$, and 3 from $n = 2$.
22. Increases; increases.
23. b
24. d
25. The potato contains water, whereas the ceramic plate does not.
26. **L**ight **a**mplification by **s**timulated **e**mission of **r**adiation
27. Laser light is monochromatic, coherent, and has amplified intensity.
28. The intensity of laser light can cause eye damage.
29. X-rays are called "braking rays" because they are formed when high-speed electrons are stopped as they hit a metal plate.

30. The X-rays are emitted by decelerating electrons. The greater the voltage, the higher the energy of the electrons, and thus the shorter the cutoff wavelength.
31. In fluorescence, the light stops coming from the sample at the instant the exciting source is removed. In phosphorescence, the light continues to be emitted from the sample for a period of time. The electron transitions to lower levels occur instantaneously in fluorescence but take more time in phosphorescence.
32. b
33. A moving particle has a wave associated with it called a matter wave, or de Broglie wave. The associated wavelength is significant only for atoms and subatomic particles.
34. Davisson and Germer showed that a beam of electrons undergoes diffraction, which is a wave phenomenon.
35. The electron microscope.
36. a
37. b
38. It gives the probability of finding an electron at that point.
39. An *orbit* is a definite electron path at a given radius from the nucleus. An *orbital* is a region in space about the nucleus in which there is a high probability of finding an electron.
40. The Bohr model did not give accurate results with multielectron atoms.
41. Only n; both n and l.
42. Each electron in an atom must have its own unique set of four quantum numbers.
43. A $1p$ subshell is impossible, because for $n = 1$, l can only be 0 (for a p subshell, l would have to be 1).
44. "spins" (or m_s values)
45. 2, 6, 10, 14
46. $4s^1$
47. It is the subshell of lowest energy in atoms of every element, and so it is the first to fill.
48. c
49. b
50. If there are limits on the precision of measurement of the position and velocity of a particle, it is impossible to predict with certainty its position and velocity at a future time.
51. Only particles of atomic and subatomic size would have their velocities significantly changed by being hit by photons when locating their positions.

ANSWERS TO CRITICAL THINKING QUESTIONS

1. Microwaves could damage the person operating the oven, because a person's body contains water and could thus be "cooked" by stray microwaves.
2. Of all the colors of visible light, the photons of red light are lowest in energy and thus least likely to damage photographic film.
3. Because TV picture tubes operate by playing a beam of electrons on a phosphor-coated screen, some X-rays are formed as the electrons are decelerated.

4. Practically no light would get through, because the four wavelengths emitted by the gas-discharge tube of hydrogen are exactly those wavelengths that the cool hydrogen gas would absorb. As the excited cool hydrogen gas re-emits the four wavelengths in all directions, a little radiation would be emitted in the direction of the screen, and so a very faint line emission spectrum might be seen.

5. The black cat represents the classical view, because that cat could sleep at any level on the ramp. The orange cat, on the other hand, can sleep only on discrete levels (the steps), and thus its positions are quantized.

6. "Black light" is the popular term for the longer-wavelength portion of the ultraviolet region. You would explain that the photons of ultraviolet light were causing electrons in the atoms of the paint to be excited to upper energy levels. Some of the electrons return to their ground states in steps, emitting some photons lower in energy than the UV photons that excited them. Some of the lower-energy photons emitted are in the visible region, making the picture glow. We call the phenomenon fluorescence.

ANSWERS TO EXERCISES

1. $E = hf = (6.63 \times 10^{-34}\text{ J-s})[(5.45 \times 10^{14}/\text{s})] = \underline{3.61 \times 10^{-19}\text{ J}}$

2. $E = hf = (6.63 \times 10^{-34}\text{ J-s})[(5.00 \times 10^{14}/\text{s})] = \underline{3.32 \times 10^{-19}\text{ J}}$

3. (a) $E = hf; f = \dfrac{E}{h} = \dfrac{6.63 \times 10^{-19}\text{ J}}{6.63 \times 10^{-34}\text{ J-s}} = \underline{1.00 \times 10^{15}\text{ Hz}}$

   (b) $\lambda = \dfrac{c}{f} = \dfrac{3.00 \times 10^{8}\text{ m/s}}{1.00 \times 10^{15}\text{ 1/s}} = 3.00 \times 10^{-7}\text{ m} = 300 \times 10^{-9}\text{ m} = \underline{300\text{ nm}}$

4. (a) $E = hf; f = \dfrac{E}{h} = \dfrac{1.66 \times 10^{-19}\text{ J}}{6.63 \times 10^{-34}\text{ J-s}} = \underline{2.50 \times 10^{14}\text{ Hz}}$

   (b) $\lambda = \dfrac{c}{f} = \dfrac{3.00 \times 10^{8}\text{ m/s}}{2.50 \times 10^{14}\text{ 1/s}} = 1.20 \times 10^{-6}\text{ m} = 1200 \times 10^{-9}\text{ m} = \underline{1200\text{ nm}}$

5. $r_n = 0.053\text{ nm} \times n^2 = 0.053\text{ nm} \times 9 = \underline{0.477\text{ nm}}$

6. $r_n = 0.053\text{ nm} \times n^2 = 0.053\text{ nm} \times 16 = \underline{0.848\text{ nm}}$

7. $E_n = \dfrac{-13.60}{n^2}\text{ eV} = \dfrac{-13.60}{3^2}\text{ eV} = \underline{-1.51\text{ eV}}$

8. $E_n = \dfrac{-13.60}{n^2}\text{ eV} = \dfrac{-13.60}{4^2}\text{ eV} = \underline{-0.850\text{ eV}}$

9. $E_{photon} = E_{n_i} - E_{n_f} = -0.85\text{ eV} - (-3.40\text{ eV}) = \underline{2.55\text{ eV}}$

10. $E_{photon} = E_{n_i} - E_{n_f} = -13.60\text{ eV} - (-0.85\text{ eV}) = \underline{-12.75\text{ eV}}$ (the negative indicates photon absorption)

11. $\lambda = \dfrac{h}{mv} = \dfrac{6.63 \times 10^{-34}\text{ J-s}}{(0.50\text{ kg})(26\text{ m/s})} = 0.51 \times 10^{34}\text{ m} = \underline{5.1 \times 10^{-35}\text{ m}}$

12. $\lambda = \dfrac{h}{mv} = \dfrac{6.63 \times 10^{-34}\text{ J-s}}{(6.0 \times 10^{24}\text{ kg})(3.0 \times 10^{4}\text{ m/s})} = 0.37 \times 10^{-62}\text{ m} = \underline{3.7 \times 10^{-63}\text{ m}}$

13. l can have values from 0 up to and including $n - 1$, so for $n = 3$, l can be 0, 1, and 2. These l values correspond to subshells designated *s*, *p*, and *d*.

14. l can have values from 0 up to and including $n - 1$, so for $n = 4$, l can be 0, 1, 2, and 3. These l values correspond to subshells designated *s*, *p*, *d*, and *f*.

15. (a) $1s^2 2s^2 2p^2$
    (b) $1s^2 2s^2 2p^6 3s^2 3p^3$
    (c) $1s^2 2s^2 2p^6 3s^2 3p^6 4s^2$
16. (a) $1s^2 2s^2 2p^4$
    (b) $1s^2 2s^2 2p^6 3s^2 3p^1$
    (c) $1s^2 2s^2 2p^6 3s^2 3p^6 4s^1$
17. (a) 2 electrons in the first shell, 8 in the second, 8 in the third, 2 in the fourth.
18. 2 electrons in the first shell, 8 electrons in the second shell, 3 electrons in the third shell.

ANSWERS TO RELEVANCE QUESTIONS

9.1. The basic quantum unit of our money would be the penny. No smaller unit of money is available, and so you never see something priced at, say, $1.985.

9.2. Mercury vapor or sodium vapor lights give off only certain wavelengths of light (those in their line emission spectra). Thus only certain colors are available to be absorbed or reflected from the variously colored cars. The reflected light by which we see the cars thus has a different overall composition (color) from reflected sunlight or reflected light from an incandescent bulb.

9.3. Laser light, such as from a laser "pointer," might present some problems for studying, because the beam does not spread out and also might be of dangerous intensity.

ANSWERS TO STUDY GUIDE QUIZ

Multiple-Choice Questions

1. c	2. a	3. b	4. c	5. d
6. c	7. c	8. a	9. d	10. c

Short-Answer Questions

1. Albert Einstein developed a complete explanation of the photoelectric effect using the idea that light comes in bundles of energy called photons and that the energy of a photon is equal to Planck's constant times the frequency of the light. The quantum theory describing the behavior of light was first presented by Max Planck.

2. The ultraviolet catastrophe referred to the observations of the intensity of emitted energy from hot matter, which could have increased rapidly as the frequency increased. This effect was not observed experimentally and thus presented a very large problem (a catastrophe) for those trying to explain these observations using the classical wave theory. A new and more far-reaching theory was needed to explain these experimental data, and this was developed by Max Planck in 1900.

3. Niels Bohr had developed a theory that assigned specific energy levels within the hydrogen atom, in which electrons had to exist for the atom to be stable. Bohr suggested that the lines in the light spectra of hot gases were the result of energy given off (emitted) when an electron "jumped" downward in energy from a higher allowed energy orbit to a lower allowed orbit. This explained the emission of energy at specific frequencies that corresponded to the colored lines observed in the spectrum, and hence these lines have come to be shown as a line emission spectrum.

4. Any energy level above the ground state is referred to as an excited state, and so all three higher electron orbits go by this designation. In particular, the first excited state has the principal quantum number $n = 2$, the second excited state has $n = 3$, and the third excited state has $n = 4$.

5. Laser light is emitted in phase, in one specific direction, and all at the same frequency. A normal incandescent bulb emits light in random phase, in a spherical pattern surrounding the bulb, and at many different frequencies.

6. The dual nature of matter indicated that moving matter (usually particles) shows not only particle characteristics but also wave characteristics. This is, however, generally observed only for individual atoms and subatomic particles and is not easily detected for ordinary-sized moving bodies such as cars or even baseballs.

7. The matter wave equation was developed by Erwin Schrödinger using a complex mathematical format, the most striking feature of which is that the square of this wave function is an indication of the probability of finding the electron in the hydrogen atom at a specific distance from the nucleus. This theory gives rise to the modern quantum mechanical theory of the atom, in which the orbits of the electrons are now specified as electron-clouds rather than having specific, fixed radii.

8. Heisenberg's uncertainty principle says that it is not possible to precisely measure both the position and the velocity of a subatomic particle at the same time.

9. The four quantum numbers needed to completely specify the energy of an electron are:
   (1) n, the principal quantum number—indicates the primary energy of the electron orbit.
   (2) ℓ, the orbital quantum number—associated with both the energy and the angular momentum of the electron.
   (3) m_ℓ, the magnetic quantum number—gives the orientation in space of an electron orbital relative to an applied magnetic field.
   (4) m_s, the spin quantum number—which shows the clockwise of counterclockwise orbital electron spin.

10. The first shell ($n = 1$) consists of only the 1s subshell.
    The second shell ($n = 2$) consists of the 2s and 2p subshells.
    The third shell ($n = 3$) consists of the 3s, 3p, and 3d subshells.

Chapter 10

Nuclear Physics

This chapter concludes the physics section of the textbook by discussing the atomic nucleus. Because of the great impact that nuclear energy has had—and will have—on society, this chapter is extremely important. The material in this chapter is so fundamental that we recommend that it be covered in its entirety. Among the topics discussed are the structure and composition of nuclei, atomic mass, radioactive decay, half-life and radiometric dating, nuclear reactions and reactors, fission, fusion, and the biological effects of radiation. The small amount of mathematics included is kept simple. Applications of radioactivity are stressed, and two Highlights emphasize the historical aspects of the discovery of radioactivity and the building of the first nuclear bombs.

The student will need to refer frequently to the periodic table on the inside front cover of the textbook. We recommend that a wall chart of the periodic table be on display at all times in the lecture classroom, including during exams.

DEMONSTRATIONS

Students generally find a demonstration of a Geiger counter or other radiation detector to be interesting and helpful. A comparison of the penetrating power of alpha particles, beta particles, and gamma rays is instructive (a solid source set #32852 can be obtained from CENCO or another science supply company). Another demonstration involves charging an electroscope and illustrating its quick discharge with a radioactive source. Of course, use appropriate safety precautions when dealing with radiation. An old X-ray of a human or animal obtained from a doctor or veterinarian may be passed around the class. A display Chart of the Nuclides is available at moderate cost from General Electric Company, 175 Curtner Ave., Mail Code 684, San Jose, CA 95125. For additional ideas on demonstrations, refer to the publications listed in the appendix of this *Guide*.

Appropriate films and videos include *The Discovery of Radioactivity* and *The Day Tomorrow Began*. For additional ideas on appropriate audio-visual material, refer to the catalogs of the audio-visual suppliers listed in the appendix of this *Guide*.

ANSWERS TO REVIEW QUESTIONS

1. c
2. d
3. Electrons, protons, neutrons. Protons and neutrons have a mass of about 1 u, electrons have a mass of about 0.00055 u. Electrons have a charge of 1–, protons have a charge of 1+, and neutrons are neutral.

4. Nucleons.
5. Henri Becquerel in 1896.
6. The diameter of the atom is about 10,000 times that of the nucleus (10^{-10} m as compared to 10^{-14} m).
7. Over 99.9%.
8. Protons.
9. Z is the atomic number, A is the mass number, and N is the neutron number. $N = A - Z$.
10. Protium, deuterium, and tritium.
11. The mass spectrometer. The atoms are ionized and sent through a magnetic field. Because the ions of different isotopes have the same charge but different mass, they interact differently with the field and form separate beams.
12. Carbon-12, which is assigned a mass of exactly 12 u.
13. The strong nuclear force (or just *nuclear force*). It drops to zero at distances greater than about 10^{-14} m.
14. d
15. b
16. c
17. Marie and Pierre Curie.
18. $A \rightarrow B + b$.
19. ${}^{4}_{2}\text{He}$, ${}^{0}_{-1}\text{e}$, γ.
20. (a) Gamma.
    (b) Beta.
    (c) Alpha.
    (d) Gamma.
    (e) Alpha
21. Mass numbers, atomic numbers.
22. 83
23. d
24. b
25. One-fourth.
26. The half-life of ${}^{238}\text{U}$ is about 4.5 billion years, which is about the same as the age of Earth, and so about half the original amount of ${}^{238}\text{U}$ should still be present.
27. c
28. $a + A \rightarrow B + b$
29. Used in smoke detectors as a source of ionizing radiation.
30. Neutron activation analysis.
31. a
32. Neutrons.
33. The minimum amount of fissionable material necessary to sustain a chain reaction is called the critical mass. A subcritical mass would be less than that, and no chain reaction would be possible. A supercritical mass would be more than that, and an expanding chain reaction would be possible.

34. 0.7%, which must be enriched to about 3% for use in a U.S. nuclear reactor, and to 90% or more for use in nuclear weapons.
35. Enrico Fermi, at the University of Chicago in 1942.
36. The *Manhattan Project* was the name applied to the United States effort to build an atomic bomb.
37. Control rods adjust the number of neutrons available to cause fission. Moderators slow down fast neutrons from fission so that they can more effectively cause other fissions.
38. For a nuclear explosion to occur, the fissionable material must be of much greater purity than the fuel in nuclear reactors.
39. Plutonium-239 is made from uranium-238.
40. c
41. Nuclear fusion.
42. D is a deuterium nucleus, or deuteron; T is a tritium nucleus, or triton.
43. A plasma is a very hot gas of electrons and protons or other nuclei. Electric fields are used to form the plasma, and electric and magnetic fields are used to confine its charged particles.
44. Fusion has the advantages of low cost and abundance of fuel, fewer nuclear waste disposal problems, and the impossibility of a runaway accident. Its disadvantages are that it has not yet proved practical, and that fusion plants will be more costly to build and operate than fission plants.
45. Einstein. E is energy, m is mass, and c is the speed of light.
46. Exoergic, mass defect.
47. A sketch of mass defect per nucleon vs. mass number should be shown, with fission indicated by an upward-pointing arrow on the right, and fusion indicated by an upward pointing arrow on the left. See Figure 10.25.
48. b
49. 6000 mSv.
50. Somatic effects are short-term and long-term effects on the health of the recipient of the radiation. Genetic effects are defects in the recipient's subsequent offspring.
51. Natural sources include cosmic rays, radionuclides such as ^{14}C and ^{40}K in the body, and radionuclides such as ^{238}U and ^{232}Th in rocks. Artificial sources include medical X-rays, fallout, TVs, tobacco smoke, nuclear wastes, radionuclides used in medical procedures, and emissions from power plants.

ANSWERS TO CRITICAL THINKING QUESTIONS

1. Nuclides must have the same atomic number to be correctly called isotopes. The two reported have atomic numbers of 112 and 114, and thus are not isotopes in the strict sense.
2. No, you don't invest, because you know that chemical procedures cannot cause transmutation, the change of one element to another.
3. We rely on the Sun's energy, which comes from the process of nuclear fusion.
4. *China syndrome* refers to the meltdown of the core of a nuclear reactor through the Earth so that it winds up in China. This is, of course, impossible. And anyway, most nuclear reactors are in the Northern Hemisphere, as is China. The term was used as the title of a movie starring Jane Fonda.

5. The major "pro" would be that, if successful, the procedure would take the radioactive material completely from our environment. The major "cons" are that it would be expensive and there is a possibility of an accident during launch that would spread radioactive material around.
6. Some short-half-life radionuclides are formed continually by cosmic rays. Others are part of the decay series of long-lived radionuclides and so are replenished.
7. The neon-20 ion beam would be deflected more, because these ions have less mass than those of neon-22.
8. They both have 20 neutrons in their atoms.

ANSWERS TO EXERCISES

1. (a) 7, 3, 3, 4, Li
   (b) 239, 93, 93, 146, Np
   (c) 31, 15, 15, 16, P
   (d) 34, 16, 16, 18, S
2. (a) 90, 38, 38, 52, Sr
   (b) 235, 92, 92, 143, U
   (c) 11, 5, 5, 6, B
   (d) 240, 94, 94, 146, Pu
3. $0.5069 \times 78.918 \text{ u} = 40.00 \text{ u}$
   $0.4931 \times 80.916 \text{ u} = \underline{39.90 \text{ u}}$
   79.90 u
4. $0.9326 \times 38.964 \text{u} = 36.34 \text{ u}$
   $0.00012 \times 39.964 \text{ u} = 0.0048 \text{ u}$
   $0.0673 \times 40.962 \text{ u} = \underline{2.76 \text{ u}}$
   39.10 u
5. (a) γ, gamma
   (b) ${}^{228}_{88}\text{Ra}$, alpha
   (c) ${}^{0}_{-1}\text{e}$, beta
6. (a) ${}^{237}_{94}\text{Pu}$, beta
   (b) ${}^{4}_{2}\text{He}$, alpha
   (c) γ, gamma
7. (a) ${}^{228}_{88}\text{Ra} \rightarrow {}^{222}_{86}\text{Rn} + {}^{4}_{2}\text{He}$
   (b) ${}^{60}_{27}\text{Co} \rightarrow {}^{60}_{28}\text{Ni} + {}^{0}_{-1}\text{e}$
8. (a) ${}^{229}_{90}\text{Th} \rightarrow {}^{225}_{88}\text{Ra} + {}^{4}_{2}\text{He}$
   (b) ${}^{225}_{88}\text{Ra} \rightarrow {}^{225}_{89}\text{Ac} + {}^{0}_{-1}\text{e}$
9. (a) ${}^{249}_{98}\text{Cf}$ $(Z > 83)$
   (b) ${}^{76}_{33}\text{As}$ (odd-odd)
   (c) ${}^{15}_{8}\text{O}$ (fewer n than p)
   (d) ${}^{33}_{15}\text{P}$ (odd-even, and A >1.5 u from atomic mass)

10. (a) $^{111}_{47}Ag$ (odd–even, with A >1.5 u from atomic mass)
    (b) $^{17}_{9}F$ (fewer n than p)
    (c) $^{226}_{88}Ra$ (Z> 83)
    (d) $^{24}_{11}Na$ (odd-odd); nitrogen-14 is one of the four odd-odd exceptions.

11. $\frac{36\text{ h}}{6.0\text{ h/half-life}} = 6$ half-lives

12. 160 cpm → 80 cpm → 40 cpm → 20 cpm → 10 cpm → 5 cpm
    Five arrows are shown, so five half-lives are neeeded.

13. $\frac{24.3\text{ d}}{8.1\text{ d/half-life}} = \underline{3\text{ half-lives}}$
    1 → 1/2 → 1/4 → 1/8

14. $\frac{75\text{ h}}{15\text{ h/half-life}} = \underline{5\text{ half-lives}}$
    480 cpm → 240 cpm → 120 cpm → 60 cpm → 30 cpm → 15 cpm will be the activity.

15. 1→ 1/2 → 1/4 → 1/8 → 1/16
    Four arrows are shown, so four half-lives have gone by. (4 half-lives)(12.3 y/half-life) = 49 y

16. 15.3 → 7.65 → 3.83
    Two arrows are shown, so two half-lives have gone by. (2 half-lives)(5730 y/half-life) = 11,500 y

17. For radionuclide A, half-life = 22 d , because half of 40 g is 20 g, and 20 g is reached after 22 d.

18. For radionuclide B, the original amount is 32 g, half of which would be 16 g. From the graph, 16 g is left after 8 days have gone by, so the half-life = 8 d .

19. (a) $^{1}_{1}H$
    (b) $^{1}_{0}n$
    (c) $^{2}_{1}H$
    (d) $^{93}_{38}Sr$

20. (a) $^{24}_{12}Mg$
    (b) $^{28}_{13}Al$
    (c) $^{1}_{1}H$
    (d) $^{67}_{32}Ge$

21. $3\,^{1}_{0}n$

22. $^{106}_{43}Tc$

23. 3 ×4.00260 u = 12.00780 u on left
    −12.00000 u on right
    0.00780 u mass loss
    (0.00780 u)(931 meV/u) = 7.26 MeV of energy produced

24. 2,0140 u + 2,01401 u = 4,0280 u showing on the left

3.0161 u + 1,0078 u = 4.0239 u showing on the right

The difference is 0.0041 u more on the left, so mass is lost and (0.0041 u)(931 MeV/u) = 3.8 MeV of energy is produced.

ANSWERS TO RELEVANCE QUESTIONS

10.1. The term "star stuff" is appropriate because most of the atoms in our bodies were formed in the cores of stars and during supernova explosions.

10.2. Answers may vary according to the individual, but major considerations would be that coal-burning plants release more radioactivity than do nuclear plants during normal operation, and the emissions of coal-burning plants would cause much grime to collect on and in your house and could well affect adversely your respiratory system.

ANSWERS TO STUDY GUIDE QUIZ

Multiple-Choice Answers

1. b	2. d	3. c	4. d	5. d
6. d	7. a	8. a	9. d	10. b

Short-Answer Questions

1. The isotope has 8 protons and is oxygen, and so the atomic number (Z) must be 8. The neutron number (N) will be 9. The number of protons plus the number of neutrons must equal the mass number (A), which in this case will be 8 + 9 = 17. The chemical notation is then ${}^{17}_{8}O_9$.
2. The element hydrogen must have one proton, but it can have 0, 1, or 2 neutrons in its nucleus. Normal hydrogen has no neutron. Tritium is the special isotope of hydrogen that has two neutrons (deuterium has one), and so the isotope symbol for tritium is ${}^{3}_{1}H_2$, or just ${}^{3}_{1}H$.
3. All nuclear reactions follow (1) the conservation of mass number and (2) the conservation of atomic number.
4. The numbers of protons and neutrons in the nucleus of an atom affect is stability, and it has been found that the nuclides with an even number of protons and an even number of neutrons in their nuclei are the most stable. These are referred to as even-even nuclides. Practically all other stable nuclides are either even-odd or odd-even, which indicates that somehow nature does not like odd-odd combinations of protons and neutrons in nuclides.
5. Alpha decay—emits a helium nucleus ${}^{4}_{2}He$, which is also known as an alpha particle. Beta decay—emits an electron, ${}^{0}_{-1}e$, which is also known as a beta particle. Gamma decay—emits a proton of electromagnetic radiation known as a gamma ray.
6. Both heavy and light nuclides are less stable than those in the center portion of the periodic table, around the mass number of iron. This means that light nuclei can combine in fusion reactions and release energy in the process, and heavy nuclei can split apart in fission reactions and also release energy in the process.

7. If 87.5% of the radionuclide has decayed, 12.5% must be left. This means that three half-lives have passed, as shown below.
   The initial 100% decays to 50% in the first half-life.
   The remaining 50% decays to 25% in the second half-life.
   The remaining 25% decays to 12.5% in the third half-life.
8. The atomic number total = 80 + 1 = 81 on the left side of the equation. It must be the same on the right side, and so the other particle will have 81 – 79 = 2.
   The mass number = 200 + 1 = 201 on the left side of the equation. It must also equal 201 on the right side. This means that the mass number of the unknown nuclide must be 201 – 197 = 4.
   The nuclide with an atomic number of 2 and a mass number of 4 is ${}^{4}_{2}He$, and so the equation will be
   $$ {}^{1}_{1}H + {}^{200}_{80}Hg \rightarrow {}^{197}_{79}Au + {}^{4}_{2}He $$
9. When a neutron interacts with a uranium-235 nucleus, a fission reaction occurs in which two lighter nuclei plus one or more additional neutrons are produced. The newly released neutrons can then interact with another uranium-235 nucleus to produce another fission reaction, and this process can continue as a chain reaction. If suitable control (neutron absorption) rods are used, the interactions can be controlled so that only one new fission is produced for each previous fission reaction, and a controlled chain reaction will be the result.
10. A meltdown can occur when a nuclear chain reaction is not controlled and the chain reaction accelerates rapidly, producing large enough amounts of heat to actually "melt down" the structure of the nuclear reactor core. This can lead to the release of radioactive material into the atmosphere or into the groundwater, resulting in a large-scale environmental problem.

Chapter 11

The Chemical Elements

This chapter begins the five-chapter chemistry section by discussing the chemical classification of matter, the elements, the periodic table, and the basics of compound nomenclature. For simplicity, our discussion of chemistry will involve primarily the representative (A-Group) elements.

Much of the material discussed in this chapter is necessary for dealing successfully with the next four chapters. We recommend that the student immediately learn the names and symbols of the 44 elements in Table 11.3 of the textbook, the names and formulas of the eight polyatomic ions listed in Table 11.7, and the names and formulas of ammonia, methane, nitrous oxide, nitric oxide, and the six common acids mentioned in Section 11.5. Probably the most efficient and effective way of doing this is to make and use flashcards.

The student will need to refer frequently to the periodic table on the inside front cover of the textbook. We recommend that a wall chart of the periodic table be on display in the lecture classroom at all times, including during exams. The material in this chapter is so fundamental that we recommend that it be covered in its entirety.

DEMONSTRATIONS

When discussing the differences between compounds and mixtures, it is useful to pass around two Erlenmeyer flasks, one containing the compound zinc sulfide and the other containing a heterogeneous mixture of powdered sulfur and mossy zinc. For additional ideas on demonstrations, refer to the publications listed in the Teaching Aids Section of this *Guide.*

A historical periodic poster that pictures the elements and gives some of their histories, helps the students to visualize many of the descriptions given in the textbook. It can be obtained from places such as Fisher EMD (S45527-1) for about $25, and might be posted on the bulletin board in the classroom. Perhaps the best basic chemistry film ever made is *The Periodic Table*, available from Media Guild. For additional ideas on appropriate audio-visual material, refer to the catalogs of the audio-visual suppliers listed in the Teaching Aids Section of this *Guide*.

ANSWERS TO REVIEW QUESTIONS

1. *Physical chemistry* applies the theories of physics (especially thermodynamics) to the study of chemical reactions.

   Analytical chemistry deals with the identification of substances present in a material and how much of each substance is present.

   Organic chemistry is the study of compounds that contain carbon.

Inorganic chemistry is the study of compounds that do not contain carbon.

Biochemistry studies the chemical reactions that occur in living organisms.

2. c
3. Illustrations b and d show different atoms and/or molecules and thus represent mixtures. Illustration e shows identical molecules composed of the same two elements and so represents a compound. Illustrations a and c show identical atoms and identical diatomic molecules of atoms of the same element, respectively, and thus represent elements. Illustration c shows diatomic molecules of an element, whereas e shows diatomic molecules of a compound.
4. A *pure substance* is a type of matter in which all samples have fixed composition and identical properties, whereas a *mixture* is a type of matter composed of varying proportions of two or more substances that are just physically mixed, not chemically combined. Thus samples of a mixture can have variable composition and properties.

   An *element* is a pure substance in which all the atoms have the same number of protons; elements cannot be further broken down by chemical processes. A *compound* is a substance composed of two or more elements chemically combined in a definite, fixed proportion by mass.
5. Mixtures can be separated into, and prepared from, pure substances by physical processes.

   Compounds can be prepared from, and decomposed into, elements by chemical processes only.
6. An appropriate set of examples would be: bronze, salt water, air.
7. a
8. c
9. The Greeks, about 600 to 200 B.C.
10. Aristotle thought that all matter on Earth was composed of four elements: earth, air, fire, and water.
11. Humphry Davy set a record for the discovery of elements (six) with the new invention called the battery.
12. At present, 112 elements are known, about 90 of which occur naturally on Earth. Technetium was the first artificial element.
13. J. J. Berzelius first conceived the symbol notation used today for elements. Some of the symbols are taken from the Latin name and so bear little resemblance to the English name of the element, making them less easy to learn.
14. d
15. b
16. (a) oxygen and silicon.
    (b) iron and nickel.
    (c) oxygen and carbon.
    (d) nitrogen and oxygen.
    (e) hydrogen and helium.
17. d
18. c
19. b
20. Dmitri Mendeleev developed the periodic table in 1869.
21. Atomic number is now used. The periodic trends in the table are based on atomic number, which correlates more exactly with electron configuration than does atomic mass.

22. *The properties of elements are periodic functions of their atomic numbers,* which means that the elements' properties show regular trends, with similar properties recurring at regular intervals.

23. (a) Periods.
    (b) Groups.

24. The valence electrons are the ones that form chemical bonds.

25. The groupings are caused by the electron capacity of *s*, *p*, *d*, and *f* subshells, respectively.

26. Lanthanides and actinides.

27. (a) Metals usually have 1 to 3 valence electrons; nonmetals usually have 4 to 8.
    (b) Metals are good conductors of heat and electricity; nonmetals are not.
    (c) Metals are solid at room temperature (except for Hg); many nonmetals are gaseous or liquid.

28. Metallic character (a) decreases from left to right across a period, and (b) increases down a group. Cesium is the most metallic element, and fluorine is the most nonmetallic.

29. Semimetals or metalloids, located next to the staircase line in the periodic table.

30. Mercury and bromine are liquid. Hydrogen, oxygen, nitrogen, fluorine, chlorine, and the noble gases (helium, neon, argon, krypton, xenon, and radon) are gases.

31. Atomic size (a) decreases from left to right across a period, and (b) increases down a group.

32. d

33. a

34. (a) An atom is the smallest particle of an element; a molecule is an electrically neutral particle composed of two or more atoms chemically bound together.
    (b) An atom is electrically neutral; an ion has an electrical charge.
    (c) A molecule is an electrically neutral particle composed of two or more atoms; a polyatomic ion is an electrically charged particle composed of two or more atoms.

35. Because the nomenclature rule for metal—nonmetal compounds is different from the rule for nonmetal—nonmetal compounds.

36. a

37. c

38. Because they have the same number of valence electrons.

39. The noble gases are chemically unreactive. They have low boiling and melting points.

40. Fluorine and chlorine are gases, bromine is liquid, and iodine is solid. Fluorine is the most reactive element.

41. Sodium iodide is added to prevent thyroid problems due to iodine deficiency.

42. (a) Calcium phosphate.
    (b) Calcium carbonate.

43. The *Hindenburg* syndrome is the term applied to the reluctance to use hydrogen as a fuel.

ANSWERS TO CRITICAL THINKING QUESTIONS

1. Some of the chemical reactions of hydrogen are like those of an alkali metal (Group 1A), and some are like those of a halogen (Group 7A).

2. It should resemble Pb, because counting across in the periodic table shows that it would be in the same group as Pb.

3. Although homogenized milk *looks* uniform to the unaided eye, a microscope shows fat globules dispersed throughout the water. Therefore, it is not a true solution, because it is not mixed on the atomic or molecular level.
4. Alkali metals are soft and react with air and water.
5. Uup (from una, una, and pente).
6. About 0.255 nm (certainly larger than cerium's 0.235-nm radius).

ANSWERS TO EXERCISES

1. (a) Homogeneous mixture.
   (b) Compound.
   (c) Element.
   (d) Heterogeneous mixture.
2. (a) Heterogeneous mixture.
   (b) Element.
   (c) Homogeneous mixture.
   (d) Compound.
3. (a) S
   (b) Na
   (c) Al
4. (a) Fe
   (b) Rn
   (c) Ba
5. (a) Nitrogen.
   (b) Potassium.
   (c) Zinc.
6. (a) Beryllium.
   (b) Gold.
   (c) Argon.
7. (a) 6.94 u, 3, 3, 3
   (b) 197.0 u, 79, 79, 79
8. (a) 20.2 u, 10, 10, 10
   (b) 207.2 u, 82, 82, 82
9. (a) 3, 2A
   (b) 4, 2B
   (c) 5, 4A
10. (a) 6, 1A
    (b) 5, 1B
    (c) 1, 8A
11. (a) Representative, nonmetal, gas.
    (b) Transition, metal, solid.
    (c) Inner transition, metal, solid.

12. (a) Inner transition, metal, solid.
    (b) Representative, nonmetal, gas.
    (c) Transition, metal, liquid.
13. (a) 14, 4, 3
    (b) 33, 5, 4
14. (a) 20, 2, 4
    (b) 8, 6, 2
15. (a) Ca, Mn, Se
    (b) Po, Se, O
16. (a) Na, P, Cl
    (b) Br, Cl, F
17. (a) Sr, Sn, Xe
    (b) Ar, Ne, He
18. (a) Cs, K, Na
    (b) Ca, As, Br
19. (a) Kr, Br, Ca
    (b) Li, Rb, Cs
20. (a) Ne, C, Be
    (b) Si, Ge, Pb
21. (a) Calcium bromide.
    (b) Dinitrogen pentasulfide.
    (c) Zinc sulfate.
    (d) Potassium hydroxide.
    (e) Silver nitrate.
    (f) Iodine heptafluoride.
    (g) Ammonium phosphate.
    (h) Sodium phosphide.
22. (a) Aluminum carbonate.
    (b) Ammonium sulfate.
    (c) Lithium sulfide.
    (d) Sulfur trioxide.
    (e) Barium nitride.
    (f) Barium nitrate.
    (g) Silicon tetrafluoride.
    (h) Disulfur dichloride.
23. (a) Sulfuric acid.
    (b) Nitric acid.
    (c) Hydrochloric acid.
24. (a) Phosphoric acid.
    (b) Acetic acid.
    (c) Carbonic acid.
25. Li_2S, Li_3N, $LiHCO_3$
26. $Ba(NO_3)_2$, $BaCl_2$, $Ba_3(PO_4)_2$

ANSWERS TO RELEVANCE QUESTIONS

11.1. Answers depend on the individual. Common answers will be water or salt for the compound, and air or coffee for the homogeneous mixture.

11.2. It is impossible to locate an item that is not composed of chemicals.

11.3. Helium is totally unreactive, and so there is no way to make it explode in the blimp.

ANSWERS TO STUDY GUIDE QUIZ

Multiple-Choice Questions

1. b	2. b	3. c	4. a	5. a
6. b	7. c	8. c	9. b	10. d

Short-Answer Questions

1. The two basic functions of analytical chemistry are (1) to identify what substances are present in a material, and (2) to determine how much of each substance is present.

2. (a) *Dmitri Mendeleev* receives the major credit for developing the periodic tables.
   (b) *Jons Jakob Berzelius* developed the symbol notation now used for elements.

3. The most abundant element (by mass) found in:
   (a) Earth's crust is oxygen (O).
   (b) Earth's atmosphere is nitrogen (N).
   (c) living matter is oxygen (O).
   (d) the entire universe is hydrogen (H).

4. (a) Al has a larger atomic radius than B.
   (b) B has a greater ionization energy than Li.
   (c) C has a higher number of valence electrons than Be.
   (d) K has greater metallic characteristics than Na.

5. (a) Sulfur dioxide.
   (b) Aluminum sulfide.
   (c) Sodium sulfate.

6. Representative elements are those found in Groups 1A through 8A in the periodic table. In this textbook, the representative elements are colored green in Fig. 11.12 and in the periodic table inside the front cover.

7. An atom is the smallest individual unit in a sample of an element. All atoms of the same element have the same number of protons in their nuclei and thus the same atomic number.

   A molecule is an electrically neutral particle composed of two or more atoms chemically combined.

   An ion is an atom or a molecule that has gained or lost one or more electrons and therefore is no longer in an electrically neutral state.

8. Allotropes are two or more forms of the same element that have different bonding structures in the same physical phase. Carbon is the most commonly cited example of such an element, with its solid phase allotropes being diamond, graphite, and the fullerenes. Oxygen also has two allotropic forms, diatomic oxygen and ozone, in its gaseous phase.

9. The outer shell of electrons in an atom is known as the valence shell. The valence electrons are involved in forming chemical bonds, and so the number of electrons in the valence shell of the atoms that make up a particular element determines the element's chemical properties.

10. Atoms of a metal tend to lose their valence electrons during chemical reactions, whereas nonmetal atoms tend to gain or share valence electrons. Metal atoms generally have fewer valence electrons (usually 1 to 3), and these electrons are less tightly bond to their atoms than are the valence electrons of nonmetals.

Chapter 12

Chemical Bonding

This chapter outlines the early history of chemistry and discusses the law of conservation of mass, the law of definite proportions, and Dalton's atomic theory. The basic ideas of ionic, covalent, metallic, and hydrogen bonding are discussed. After the student is able to determine ionic charges from chemical formulas, the Stock system of nomenclature is introduced. The material in this chapter should be covered in its entirety.

For success in Chapters 12 to 15, the student must understand the trends of ionic charge and number of covalent bonds formed for atoms of the representative elements, as given in Tables 12.2 and 12.5, and must develop the ability to write the formulas of ionic and covalent compounds using the ideas presented in Sections 12.4 and 12.5.

DEMONSTRATIONS

Large models of NaCl, $CaCO_3$, and solid CO_2 are helpful when discussing the structure of ionic and molecular solids. These are available from sources such as Klinger. For additional ideas on demonstrations, refer to the publications listed in the Teaching Aids section of this *Guide.*

Excellent films or videos for this chapter are *The Chemical Bond and Atomic Structure*, second edition, available from Coronet; *Shapes and Polarities of Molecules*, available from Ward's Multimedia; and *The Physics and Chemistry of Water*, available from BFA. For additional ideas on appropriate audio-visual material, refer to the catalogs of the audio-visual suppliers listed in the Teaching Aids section of this *Guide.*

ANSWERS TO REVIEW QUESTIONS

1. b
2. b
3. *The law of conservation of mass* states that no detectable change in total mass occurs during a chemical reaction. For example, if a sealed vessel containing copper and oxygen has a mass of 245.00 g and is heated so that copper oxide is formed, the vessel and its contents will still have a mass of 245.00 g after reaction.
4. Lavoisier discovered the law of conservation of mass, explained the role of oxygen in combustion, established the principles for naming chemicals, wrote the first modern chemistry textbook, and introduced quantitative methods into chemistry (four of the five are requested).

5. Lack of experimentation and of gathering facts.
6. The alchemists' goals were to change common metals into gold and to discover an "elixir of life." Paracelsus emphasized the goal of preparation of medicines.
7. A.D. 500 to 1600. By having reasonable objectives and avoiding mysticism, superstition, and secrecy.
8. Joseph Priestley discovered oxygen, which he called "dephogisticated air."
9. c
10. Different samples of a pure compound always contain the same elements in the same proportion by mass. For example, if one sample of pure water contains oxygen and hydrogen in a mass ratio of 8.00 to 1.00, all other samples of pure water will have that same proportion by mass. The law was found by J. L. Proust.
11. Dry ice is solid carbon dioxide, which is very cold (–78°C) and sublimes (passes directly from the solid phase to the gaseous).
12. The reactant used completely is the limiting reactant (*A* in this example). *B* is the excess reactant.
13. d
14. Dalton postulated (1) that each element is composed of small indivisible particles called atoms, which are identical for that element but different from atoms of other elements, (2) that chemical combination is the bonding of a definite number of atoms of each of the combining elements to make one molecule of the formed compound, and (3) that no atoms are gained, lost, or changed in identity during a chemical reaction.
15. d
16. c
17. d
18. In forming compounds, atoms tend to gain, lose, or share valence electrons to achieve the electron configurations of the noble gases. That is, they tend to get eight electrons (an octet) in the outer shell (except for H, which tends to get two electrons in the outer shell, like the configuration of He). Hydrogen is an exception because it must use the first shell, which can hold only two electrons.
19. The valence electrons, those in the outermost shell of the atom.
20. Atoms with three or fewer electrons tend to lose them; atoms with four or more valence electrons tend to gain more electrons.
21. By transfer of electrons.
22. Two Na^+ ions and one O^{2-} ion.
23. The nucleus and all the inner shell electrons. The number of dots is the same as the atom's number of valence electrons.
24. A Lewis symbol shows the valence electrons of a single atom or ion. Lewis structures show the valence electrons in the molecules and ion combinations that make up compounds.
25. Metals form positive ions, whereas nonmetals form negative ions. Cations are positive ions; anions are negative.
26. *Isoelectronic* means having the same electron configuration (in this case, two electrons in the first shell and eight in the second).
27. The total electric charge must be zero, and all atoms must have noble gas configurations.
28. If it conducts electricity when molten, it is ionic.

29. The Stock system is preferred whenever a metal that can form more than one type of ion is part of the compound.
30. a
31. a
32. Covalent compounds are formed by the sharing of pairs of electrons. The sharing must occur in such a manner that each atom gets a noble gas electron configuration.
33. A single bond is one shared pair of electrons between two atoms, whereas double and triple bonds result from the sharing of two and three pairs, respectively.
34. Electronegativity increases from left to right in a period and decreases down a group. Fluorine has the highest electronegativity.
35. A polar covalent bond.
36. No; since there is only one polar bond, the polarity could not be offset by an opposing polar bond.
37. (a) Ionic compounds are always solid; covalent compounds may be solid, liquid, or gas.
    (b) Ionic compounds conduct electricity when melted; covalent compounds do not conduct.
38. Covalent. With a boiling point of –10°C, it is a gas at room temperature, and no ionic compounds are gaseous (or liquid) at room temperature.
39. b
40. *Like dissolves like*, which means that polar liquids tend to dissolve polar and ionic compounds but not covalent compounds, whereas covalent liquids tend to dissolve other covalent compounds but not polar and ionic substances.
41. A hydrogen bond is a dipole-dipole force between a hydrogen atom attached to O, F, or N in one molecule and a close O, F, or N atom in the same or a neighboring molecule.
42. b

ANSWERS TO CRITICAL THINKING QUESTIONS

1. The percentage abundance of isotopes of a given element is the same in all natural samples on Earth. This means that each element has an unvarying average mass (the atomic mass) that can be assigned to its atoms. The law of definite proportions holds whether atoms of a given element all have the same mass or have an unvarying average mass.
2. Solid methane would be denser than liquid methane, and so "methanebergs" would sink. Water is most unusual in having a lower density in its solid phase than in its liquid phase.
3. It must be flat, with bond angles of 120°, in order for the three polar bonds to offset one another and make the center of positive charge and the center of negative charge both be on the B atom.

ANSWERS TO EXERCISES

1. 68 tons + 96 tons – 36 tons = 128 tons
2. 432 g + 68 g + 32 g – 36 g = 496 g

3. (a) 12.0 u + (2 × 16.0 u) = 44.0 u
   (b) 12.0 u + (4 × 1.00 u) = 16.0 u
   (c) (3 × 23.0 u) + 31.0 u + (4 × 16.0 u) = 164.0 u

4. (a) 23.0 u + 35.5 u = 58.5 u
   (b) 23.0 u + 1.00 u + 12.0 u + (3 × 16.0 u) = 84.0 u
   (c) 40.1 u + (4 × 12.0 u) + (6 × 1.00 u) + (4 × 16.0 u) = 158.1 u

5. 100.0% – 25.5% = 74.5%

6. 100.0% – 20.2% = 79.8%

7. (a) (23.0 u/58.5 u) × 100% = 39.3% Na ; (35.5 u/58.5 u) × 100% = 60.7% Cl
   (b) 42.1% C , 6.4% H , 51.5% O (FM = 342.0 u)

8. (a) (23.0 u/84.0 u) × 100% = 27.4% Na , (1.00 u/84.0 u) × 100% = 1.2% H ;
   (12.0 u/84.0 u) × 100% = 14.3% C ; (48.0 u/84.0 u) × 100% = 57.1% O
   (b) 69.9% Fe , 30.1% O (FM = 159.6 u)

9. (a) (6.1 g/14.1 g) × 100% = 43% Mg
   (b) 14.1 g – 6.1 g = 8.0 g S , law of conservation of mass

10. (a) (7.75 g/67.68 g) × 100% = 11.5% P
    (b) 67.68 g – 7.75 g = 59.93 g Br , law of conservation of mass

11. Still 14.1 g of MgS, because 6.1 g of magnesium will react with only 8.0 g of sulfur, leaving 2.0 g of sulfur in excess; law of definite proportions.

12. Still 67.68 g of PBr_3, because 59.93 g of bromine will react with only 7.75 g of phosphorus, leaving 2.25 g of phosphorus in excess; law of definite proportions.

13. (a) 2–
    (b) 1+
    (c) 1–
    (d) 3–
    (e) 2+
    (f) 0
    (g) 0
    (h) 3+

14. (a) 1–
    (b) 2–
    (c) 3+
    (d) 0
    (e) 3–
    (f) 2+
    (g) 1+
    (h) 0

15. $Na\cdot + Na\cdot + :\ddot{\underset{\cdot}{O}}\cdot \rightarrow Na^{+}\ Na^{+}\ :\ddot{\underset{\cdot\cdot}{O}}:^{2-}$

16. $\cdot Ba\cdot + :\ddot{\underset{\cdot\cdot}{Br}}\cdot + :\ddot{\underset{\cdot\cdot}{Br}}\cdot \rightarrow Ba^{2+}\ :\ddot{\underset{\cdot\cdot}{Br}}:^{-}\ :\ddot{\underset{\cdot\cdot}{Br}}:^{-}$

17. (a) CsI
    (b) BaF_2
    (c) $Al(NO_3)_3$
    (d) Li_2S
    (e) BeO
    (f) $(NH_4)_2SO_4$

18. (a) $(NH_4)_3PO_4$
    (b) K_3N
    (c) $Sr(C_2H_3O_2)_2$
    (d) AlP
    (e) Na_2CO_3
    (f) $MgBr_2$

19. (a) Nickel(II) hydroxide
    (b) Lead(IV) chloride
    (c) Gold(III) iodide

20. (a) Iron(III) sulfide
    (b) Titanium(IV) bromide
    (c) Tin(II) nitrate

21. (a) 2
    (b) 0
    (c) 1
    (d) 3
    (e) 4

22. (a) 1
    (b) 0
    (c) 1
    (d) 4
    (e) 3

23.
```
H : N̈ : N̈ : H        H — N̈ — N̈ — H
    Ḧ   Ḧ                 |     |
                          H     H
```

24. (b)
```
H : C : : Ö :          H — C = Ö :
    Ḧ                      |
                           H
```

25. (a) HCl
    (b) NCl_3
    (c) SCl_2
    (d) CCl_4

26. (a) H_2S
    (b) N_2S_3
    (c) SBr_2
    (d) CS_2

27. (a) Covalent, two nonmetals.
    (b) Ionic, Group 1A metal and nonmetal.
    (c) Ionic, metal and polyatomic ion.
    (d) Covalent, two nonmetals.
    (e) Covalent, all nonmetals.
    (f) Ionic, two polyatomic ions.
28. (a) Ionic, active metal and nonmetal.
    (b) Covalent, two nonmetals.
    (c) Covalent, all nonmetals.
    (d) Ionic, active metal and nonmetal.
    (e) Covalent, two nonmetals.
    (f) Ionic, metal and polyatomic ion.
29. (a) Cl is higher and farther right than Te in the periodic table, and so it is the more electronegative and the arrows would point to it.
    (b) C and O are in the same period, but O is farther right and so is the more electronegative. Thus the arrows would point to the two oxygen atoms.
30. (a) Cl is higher than Br and in the same group, and so Cl is the more electronegative atom, and the dipole arrow will point toward the Cl.
    (b) N is higher than I, but I is farther right. Figure 12.12 shows that N has EN = 3.0, whereas I has EN = 2.5. So N is the more electronegative atom, and the dipole arrow in each bond will point toward the N.

ANSWERS TO RELEVANCE QUESTIONS

12.1. Individual atoms have indeed been imaged by electron microscopes, and have even been moved around as individual atoms.

12.2. The bonding in a coat hanger is metallic, and so shifting the positions of the positive ions does not cause a change in the attraction. A piece of chalk ($CaSO_4$) is held together by ionic bonds among the Ca^{2+} ions and the SO_4^{2-} ions, and so shifting the positions of the ions causes repulsions that cause the chalk to break.

ANSWERS TO STUDY GUIDE QUIZ

Multiple-Choice Questions

1. d 2. c 3. d 4. d 5. b
6. d 7. a 8. a 9. b 10. c

Short-Answer Questions

1. The law of conservation of mass states that no detectable change in the total mass occurs during a chemical reaction.
2. (a) Antoine Lavoisier.
   (b) Joseph Proust.

3. $\mathrm{H:\overset{\displaystyle H}{\underset{\displaystyle H}{\ddot{C}}}:\ddot{\underset{..}{Cl}}:}$

4. % of nickel = (mass of nickel / mass of nickel sulfide) × 100

   = (14.68 g/22.71 g) × 100 = 64.6%

5. Calculate the formula mass of CH_3Cl. FM_{CH_3Cl} = 12.0 u + (3 × 1.01 u) + 35.5 u = 50.5 u

   Then: % C = AM_C / FM_{CH_3Cl} = 12.0 u/50.5 u = 23.8%

   % H = AM_H / FM_{CH_3Cl} = (3 × 1.01 u)/50.5 u = 6.0%

   % Cl = AM_{Cl} / FM_{CH_3Cl} = 35.5 u/50.5 u = 70.3%

6. Alchemy was a pseudoscience whose main goals were (1) to change common metals into gold and (2) to find the "elixir of life" that could restore an aging human to youthful vigor. Modern chemistry uses logic and the systematic gathering of facts by observation and experimentation to achieve reasonable objectives, while avoiding mysticism, superstition, and secrecy.

7. The octet rule says that in forming compounds, atoms tend to gain, lose, or share valence electrons to achieve the electron configuration of a noble gas. This means that each atom tries to get eight electrons in its outer shell—thus the term octet. Hydrogen is the main exception because it strives for two electrons in its outer shell like the configuration of the noble gas helium.

8. The law of definite proportions says that different samples of a pure compound always contain the same elements in the same proportion by mass. For example:

   3 g of carbon is found in 11 g of carbon dioxide.

   6 g of carbon is found in 22 g of carbon dioxide.

   9 g of carbon is found in 33 g of carbon dioxide.

   In each case, the mass of carbon is the same proportion of the total mass of carbon dioxide; that is, 3 g / 11 g, or 27.3%.

9. The valence electrons are involved in both of these bonding processes, but in ionic bonding the electrons are transferred from one atom to another and the bond between the atoms is formed by electrostatic interaction between the two ions, whereas in covalent bonding the electrons are shared between the bonding atoms. Chlorine forms an ionic bond with a sodium atom by accepting a transferred election from the sodium atom. The resulting sodium and chloride ions combine to form an ionically bonded solid, sodium chloride. On the other hand, two chlorine atoms bond to each other to form diatomic chlorine molecules by the covalent bond, in which a pair of electrons is shared by both chlorine atoms.

10. A polar molecule is one that has a positive end and a negative end because of the way in which the valence electrons behave when the bond is formed. An example of a polar molecule is the HCl molecule, in which the bonding process leaves the hydrogen end of the molecule slightly positive and the chlorine end of the molecule slightly negative.

Chapter 13

Chemical Reactions

This chapter discusses what takes place during chemical reactions, the role of energy in reactions, and the basic types of reactions. An effort has been made to make the students aware of chemical reactions that occur in their everyday lives. Acids and bases are discussed, and the concepts of oxidation and reduction are explained. We recommend that all sections of this chapter be covered.

The completing and balancing of equations that illustrate a few basic reaction types is the most important skill developed by the student in this chapter. Information and skills attained in Chapters 11 and 12 will be used extensively.

DEMONSTRATIONS

Any demonstrations of chemical reactions will be valuable aids to student understanding and appreciation. Examples include the combination reaction when magnesium burns, a combustion reaction involving burning ethanol, an acid–carbonate reaction such as $CaCO_3$ and HCl, a single-replacement reaction such as placing a zinc strip in aqueous $CuSO_4$ or (carefully) demonstrating the reaction of a small piece of Zn with dilute hydrochloric acid. Another good demonstration would be the MnO_2-catalyzed decomposition of $KClO_3$, but take appropriate safety precautions including a safety shield.

An excellent demonstration of an exothermic reaction is to place $KMnO_4$ crystals to a height of about one inch in a 16 × 150 mm test tube, place the test tube in an Erlenmeyer flask, and add about seven drops of glycerol. After a few seconds, smoke will start to rise, and very shortly a flame will shoot up in the tube. It is advisable to use a safety screen.

The difference between a physical change and a chemical change can be demonstrated by mixing powdered sulfur and iron filings. That this is only a physical change is shown by using a magnet to separate the iron filings from the sulfur. Each substance keeps its own characteristic properties, and no chemical combination has occurred. If the mixture is placed in a test tube and heated, the iron and sulfur combine to give FeS. Now a magnet will not separate the iron from the sulfur. For additional ideas on demonstrations, refer to the publications listed in the Teaching Aids section at the end of this *Guide.*

Combustion: An Introduction to Chemical Change, available from Coronet, is an excellent film or video to show at the beginning of this chapter. *Chemical Change and Temperature,* available from BFA, also fits the chapter well. For additional ideas on appropriate audio-visual material, refer to the catalogs of the audio-visual suppliers listed in the Teaching Aids section at the end of this *Guide*.

ANSWERS TO REVIEW QUESTIONS

1. The reactants are carbon dioxide and water. The products are oxygen and glucose, the catalyst is chlorophyll, and the energy source is the Sun.
2. a
3. d
4. That it is colorless, odorless, a gas, and has a very low density are all physical properties. That it reacts with oxygen to form water and with oils to form fats are chemical properties.
5. Reactants disappear, new substances (products) appear, energy is released or absorbed.
6. A reversible reaction, which eventually reaches equilibrium, where acetic acid and ethyl alcohol molecules are reacting to form ethyl acetate and water molecules at the same rate that ethyl acetate and water molecules are reacting to form acetic acid and ethyl alcohol.

   Ethyl alcohol + acetic acid $\rightleftarrows$ ethyl acetate + water
7. c
8. Some of the original chemical bonds are broken and new bonds are formed to give different chemical structures. No atoms are changed in identity.
9. Subscripts, such as the 2 in H_2O, cannot be changed, but coefficients, such as the 3 in 3 H_2O, can be changed.
10. Do not start with oxygen, because it is present in two places on the product side.
11. Two Pb atoms, four N atoms, twelve O atoms, and four nitrate ions are indicated.
12. (a) Fractional coefficients are generally not used.

    (b) The smallest set of whole number has not been used; 4, 2, 4 should be changed to 2, 1, 2.

    (c) Hydrogen should be written H_2, not H, on the product side. (The correct set of coefficients would then become 2, 2, 2, 1.)
13. (a) $A + B \rightarrow AB$

    (b) $AB \rightarrow A + B$
14. b
15. Exothermic reactions, such as the burning of methane, liberate heat to the surroundings. Endothermic reactions, such as the changing of ozone to oxygen, absorb energy from the surroundings.
16. Energy.
17. For reaction to occur, the collision must occur at the proper places in the molecules (proper orientation) and must have at least enough energy to break bonds (the activation energy).
18. Temperature, concentration, state of subdivision, and catalyst.
19. (a) An increase in temperature increases the frequency of collisions and their average energy.

    (b) An increase in reactant concentration means that collisions will be more frequent.
20. Carbon dioxide and water are the products, and oxygen is the other reactant. If O_2 is in short supply, carbon monoxide and carbon are also products.
21. Powdered zinc and powdered sulfur are in finer states of subdivision than the lumps, and so more reactant surface is available for collisions.
22. A catalyst speeds up the rate of the reaction by providing a new pathway with lower activation energy.
23. The *-ase* endings indicate that both are enzymes.

24. Aqueous, or water, solution; solid; liquid; gas; reacts to give; reversible reaction; catalyst.

25. b

26. c

27. Acids: Conduct electricity, turn blue litmus red, taste sour, react with bases to neutralize their properties, react with active metals to liberate H_2.

    Bases: Conduct electricity, change litmus from red to blue, react with acids to neutralize their properties.

28. Arrhenius acids are substances that give hydrogen ions, H^+ (or hydronium ions, H_3O^+), in water.

    Arrhenius bases give hydroxide ions, OH^-, in water.

29. (a) Hydrochloric acid, HCl.
    (b) Magnesium hydroxide, $Mg(OH)_2$.
    (c) Sodium hydroxide, NaOH.
    (d) Sodium bicarbonate (sodium hydrogen carbonate), $NaHCO_3$.
    (e) 5% acetic acid, $HC_2H_3O_2$.

30. Water and a salt. The most common mistake is writing the wrong formula for the salt.

31. Hydrates.

32. Water, carbon dioxide, and a salt.

33. $AB + CD \rightarrow AD + CB$

34. Particles of solid form in the mixture of the two clear solutions as the positive ions of one solution combine with the negative ions of the other to form an insoluble salt.

35. c

36. a

37. Oxidation is a gain of oxygen or a loss of electrons. Reduction is a loss of oxygen or a gain of electrons.

38. $A + BC \rightarrow AC + B$

39. Hydrogen gas and a salt. This is a single-replacement reaction.

40. d

41. *Electrochemistry* is the study of chemical reactions that involve the consumption or production of electric current. In electrolytic cells, a current produces a reaction; in voltaic cells, a reaction produces a current.

ANSWERS TO CRITICAL THINKING QUESTIONS

1. Charcoal burned indoors would liberate a dangerous amount of carbon monoxide gas.
2. The citric acid dissolves and ionizes, and the hydrogen ions react with the sodium hydrogen carbonate to liberate CO_2 gas—the bubbles.
3. Storage in the refrigerator at low temperatures slows down chemical reactions (and bacterial growth) in the food.
4. In the body, enzymes catalyze the reactions; thus activation energy requirements are lower.
5. Salt dissolves to liberate ions, which conduct electricity. Dissolved sugar remains as molecules, which do not conduct.

6. Auto batteries contain a lot of lead, which is very dense. They also contain highly corrosive sulfuric acid.

7. Gold and silver are not very active, but sodium and magnesium are active and form compounds rather than remaining in the elemental state.

ANSWERS TO EXERCISES

1. (a) Physical
   (b) Chemical
   (c) Chemical
2. (a) Physical
   (b) Chemical
   (c) Physical
3. (a) 1, 1, 1, 2
   (b) 4, 3, 2
   (c) 5, 3, 1, 3
   (d) 1, 2, 1, 1
   (e) 2, 2, 1
   (f) 2, 15, 12, 6
4. (a) 2, 1, 2
   (b) 4, 3, 2, 6
   (c) 2, 13, 8, 10
   (d) 2, 2, 4, 1
   (e) 8, 3, 4, 9
   (f) 1, 1, 1, 2
5. (b) Combination.
   (e) Decomposition.
   (f) Combustion.
6. (a) Combination.
   (c) Hydrocarbon combustion.
   (d) Decomposition.
7. (a) $N_2 + 3\,H_2 \rightarrow 2\,NH_3$
   (b) $2\,KCl \rightarrow 2\,K + Cl_2$
8. (a) $2\,Al + 3\,Br_2 \rightarrow 2\,AlBr_3$
   (b) $MgCO_3 \rightarrow MgO + CO_2$
9. 12 s/2 = 6 s. Each 10°C rise in temperature approximately doubles the reaction rate. This is due to the increased temperature speeding up the molecules so that their collisions are more frequent and energetic.
10. 12 s/4 = 3 s. Each 10 C° rise in temperature approximately doubles the reaction rate, and this is a 20 C° rise. This is due to the increased temperature speeding up the molecules so that their collisions are more frequent and energetic.
11. $C_3H_8 + 5\,O_2 \rightarrow 4\,H_2O + 3\,CO_2$
12. $2\,C_2H_6 + 7\,O_2 \rightarrow 4\,CO_2 + 6\,H_2O$

13. (a) $HNO_3 + KOH \rightarrow H_2O + KNO_3$
    (b) $2\ HC_2H_3O_2 + K_2CO_3 \rightarrow H_2O + CO_2 + 2\ KC_2H_3O_2$
    (c) $H_3PO_4 + 3\ NaOH \rightarrow 3\ H_2O + Na_3PO_4$
    (d) $H_2SO_4 + CaCO_3 \rightarrow H_2O + CO_2 + CaSO_4$

14. (a) $2\ HCl + Ba(OH)_2 \rightarrow 2\ H_2O + BaCl_2$
    (b) $6\ HCl + Al_2(CO_3)_3 \rightarrow 3\ H_2O + 3\ CO_2 + 2\ AlCl_3$
    (c) $H_3PO_4 + 3\ LiHCO_3 \rightarrow 3\ H_2O + 3\ CO_2 + Li_3PO_4$
    (d) $2\ Al(OH)_3 + 3\ H_2SO_4 \rightarrow 6\ H_2O + Al_2(SO_4)_3$

15. (a) $AgNO_3(aq) + HCl(aq) \rightarrow AgCl(s) + HNO_3(aq)$
    (b) $Ba(C_2H_3O_2)_2(aq) + K_2CO_3(aq) \rightarrow BaCO_3(s) + 2\ KC_2H_3O_2(aq)$

16. (a) $Na_2CO_3(aq) + Pb(NO_3)_2(aq) \rightarrow PbCO_3(s) + 2\ NaNO_3(aq)$
    (b) $2\ K_3PO_4(aq) + 3\ CuSO_4(aq) \rightarrow Cu_3(PO_4)_2(s) + 3\ K_2SO_4(aq)$

17. (a) Will not.
    (b) Will.
    (c) Will.

18. (a) Will (Zn is above Fe).
    (b) Will not (Pb is below Fe).
    (c) Will not (Ag is below H).

19. (a) $Ni + Hg(NO_3)_2(aq) \rightarrow Hg + Ni(NO_3)_2(aq)$
    (b) $Zn + H_2SO_4(aq) \rightarrow H_2 + ZnSO_4(aq)$

20. (a) $Mg + 2\ HCl \rightarrow H_2 + MgCl_2$
    (b) $2\ Al + 3\ FeSO_4(aq) \rightarrow 3\ Fe + Al_2(SO_4)_3(aq)$

ANSWERS TO RELEVANCE QUESTIONS

13.1. "Fly equilibrium" would be attained when as many flies come in the door as go out in each unit of time. It does *not* mean there are as many flies inside as outside.

13.2. Blowing on the hot briquettes supplies a higher concentration of oxygen to the surface of the briquette.

13.3. Batteries run down because the original reactants slowly are changed to products and electrons no longer can flow.

ANSWERS TO STUDY GUIDE QUIZ

Multiple-Choice Questions

1. b	2. c	3. a	4. c	5. b
6. b	7. c	8. a	9. c	10. c

Short-Answer Questions

1. A chemical reaction is a change that alters the chemical composition of a substance and hence forms one or more new substances.

2. (a) $2\ HNO_3 + Ca(OH)_2 \rightarrow Ca(NO_3)_2 + 2\ H_2O$
   (b) $2\ HCl + Ba(OH)_2 \rightarrow BaCl_2 + 2\ H_2O$

3. $Li_2 + F_2 \rightarrow 2\, LiF$ (Note: Li and F are diatomic gases.)
4. $Zn + CuSO_4(aq) \rightarrow ZnSO_4(aq) + Cu$
5. $C_3H_8 + 5\, O_2 \rightarrow 3\, CO_2 + 4\, H_2O$
6. The condition in which the net number of reactant and product molecules remains the same is called equilibrium. Equilibrium is a dynamic process in which the reactants are interacting to form products at the same rate at which the products are interacting to form reactants.
7. A reaction is considered to be exothermic when a net release of energy to the surroundings occurs during the chemical reaction. This energy release is generally in the form of heat.
8. An electrolytic cell is one in which the chemical process called electrolysis takes place. It requires the flow of direct current from a battery or other electrical source, and this current is consumed to produce the electrolytic chemical reaction.
9. Chemical reactions generally require energy input. This is called the activation energy. Heat often provides this input energy, and the higher the temperature, the higher the kinetic energy of the reactant atoms or molecules and he faster the chemical reaction will begin and progress.
10. In some reactions the activation energy for the reaction can be lowered by adding a catalyst. Catalysts can provide a surface on which the reactants can concentrate, or they can provide new reaction pathways that tend to lower the activation energy and speed up the reaction. Catalysts are not consumed in a reaction, but they are involved; and through the alternative interaction paths that they provide, the overall reaction is greatly speeded up.

Chapter 14

Moles, Solutions, and Gases

This chapter deals primarily with such quantitative aspects of chemistry as mole relationships, mass–mass stoichiometry problems, solubility and molarity, pH and titration problems, and the gas laws. The usefulness of paying close attention to *units* should be emphasized by the instructor.

Molarity and titration problems may be omitted at the discretion of the instructor. The gas law section of the chapter has been simplified from the previous edition.

DEMONSTRATIONS

The demonstration of a titration using phenolphthalein as an indicator would be helpful to the student, unless such an experience is available in lab.

The demonstration of the gaseous diffusion of NH_3 and HCl would be valuable. Use a 2-ft length of 8-mm glass tubing. Use two Q-tips, one dipped in 12 *M* HCl (careful) and the other in 15 *M* NH_4OH (careful). Place the Q-tips in opposite ends of the glass tube (use a paper towel under each end to catch drips). About five or six minutes from the time of insertion, a small, white ring of NH_4Cl should form (closer to the HCl end because HCl molecules are more massive and thus slower than NH_3 molecules). For additional ideas on demonstrations, refer to the publications listed in the appendix of this *Guide*.

Equilibrium, available from Ward's Multimedia, fits best in this chapter rather than in Chapter 13, because much of the discussion in the film is about solutions. Other good films or videos for this chapter are *Molecular Motions* and *Gas Pressure and Molecular Collisions*, both available from Ward's Multimedia. For additional ideas on appropriate audio-visual material, refer to the catalogs of the audio-visual suppliers listed in the Teaching Aids section at the end of this *Guide*.

ANSWERS TO REVIEW QUESTIONS

1. d
2. b
3. 12; 6.02×10^{23}
4. A mole of any substance is a mass equal to the formula mass in grams (the molar mass), and is also 6.02×10^{23} formula units.
5. Avogadro, an Italian, about 1811.
6. c
7. b

8. NaBr (sodium bromide).
9. Water is the solvent and sugar the solute. The solution is saturated when no more sugar will dissolve.
10. A supersaturated solution is one that contains more than the normal maximum amount of dissolved solute. It could be formed by preparing an unsaturated solution at high temperature and carefully cooling it until the saturation point is passed. If no nucleation sites are present, crystallization may not occur.
11. Introducing seed crystals to act as nucleation sites into air that is supersaturated with water vapor.
12. Antifreeze lowers the freezing point of the water in the radiator.
13. Carbon dioxide, which escapes into the air from an opened bottle as the pressure is released and the solution warms.
14. The dissolved salt lowers the freezing point of the water. Also, some salts liberate heat as they dissolve, and thus help melt the ice.
15. To find molarity, the number of moles of solute must be divided by the number of liters of *solution*; not by the number of liters of *solvent.*
16. a
17. b
18. c
19. They are both 1.0×10^{-7} M.
20. 1000; each pH unit is a change of 10-fold, and $10 \times 10 \times 10 = 1000$.
21. At pH 10, the solution is basic, and litmus will be blue and phenolphthalein will be pink. At pH 4, the solution is acidic, and litmus will be red and phenolphthalein will be colorless.
22. A pH meter is used. The one in Fig. 14.14 shows a pH of 12.000, which indicates a basic solution.
23. Titration; buret.
24. (a) 1
    (b) 2
25. d
26. a
27. c
28. A gas consists of molecules moving independently in all directions at high speeds; the higher the temperature, the higher the average speed. The molecules collide with one another and with the walls of the container. The distance between molecules is, on average, large compared to the size of the molecules themselves.
29. In an ideal gas, the molecules have no size at all and interact only by collision. A real gas stops behaving as an ideal gas when it is under so much pressure that the space between its molecules becomes small relative to the size of the molecules themselves, or when the temperature drops to the point at which attractions among the molecules can be significant.
30. Pressure is the force per unit area; $p = F/A$. Gas pressure is caused by collisions of the molecules with the container walls.
31. The average speed of gas molecules is directly proportional to the Kelvin temperature.
32. At Oak Ridge, Tennessee, by gaseous diffusion of $^{235}UF_6$ and $^{238}UF_6$.

33. (a) $p \propto n$; the more molecules, the more collisions.
    (b) $p \propto T$; the faster the molecules, the more collisions and the harder the collisions.
    (c) $p \propto 1/V$; the greater the volume, the greater the area over which the collisions occur.
34. The pressure builds up rapidly as a closed container is heated, because the volume and number of molecules remain constant. An explosion can result.
35. The basketball stays inflated because the molecules remain in motion in all directions and keep exerting pressure.

ANSWERS TO CRITICAL THINKING QUESTIONS

1. Snowshoes increase the area over which the person's force (weight) is distributed. Thus the pressure is decreased, and the person is less likely to sink into the snow.
2. The tire temperature increases as the car is driven, and so the pressure must increase, because V and n remain essentially constant.
3. Tea contains an acid-base indicator that changes color as the solution becomes acidic.
4. Molecules of hot gas are produced by combustion, and they fill the balloon. These faster-moving molecules can exert the same pressure as cold air without having as many molecules per unit of volume. Thus hot air is less dense than cold air, and the balloon can rise in the surrounding cooler air.
5. For each mole of charge, the passage of 96,487 C is required. Dividing 192,970 C by 96,487 C/mol of charge gives 2.00 mol of charge. Therefore, the charge on the metal ion is 2+.
6. The basketball would be somewhat deflated. The cold weather would have decreased the average speed of the gas molecules inside, thus decreasing the pressure inside.

ANSWERS TO EXERCISES

1. (a) $12.0\text{ g} + (4 \times 1.00\text{ g}) = \underline{16.0\text{ g/mol}}$
   (b) $17.0\text{ g} + (3 \times 1.00\text{ g}) = \underline{17.0\text{ g/mol}}$
2. (a) $39.1\text{ g} + 79.9\text{ g} = \underline{119.0\text{ g/mol}}$
   (b) $23.0\text{ g} + 35.5\text{ g} = \underline{58.5\text{ g/mol}}$
3. (a) 27.0, 6.02×10^{23}
   (b) 1.00, 6.02×10^{23}
   (c) 1.00, 16.0
   (d) 34.0, 12.0×10^{23}
   (e) 3.00, 176
   (f) 3.00, 18.0×10^{23}
4. (a) 1.00, 6.02×10^{23}
   (b) 2.00, 6.02×10^{23}
   (c) 2.00, 36.0
   (d) 2.00, 12.0×10^{23}
   (e) 357, 18.0×10^{23}
   (f) 1.00, 78.1
5. $55.8\ g_{Fe} \times \frac{2.00\ g_{H_2}}{55.8\ g_{Fe}} = \underline{2.00\ g_{H_2}}$

6. $46\ g_{NO_2} \times \frac{126\ g_{HNO_3}}{138\ g_{NO_2}} = \underline{42\ g_{HNO_3}}$

7. $63.0\ g_{HNO_3} \times \frac{106\ g_{Na_2CO_3}}{126\ g_{HNO_3}} = \underline{53.0\ g_{Na_2CO_3}}$

8. $98.1\ g_{H_2SO_4} \times \frac{196\ g_{H_3PO_4}}{294.3\ g_{H_2SO_4}} = \underline{65.3\ g_{H_3PO_4}}$

9. About 65 g/100 g H_2O

10. About 170 g/100 g H_2O (interpolated from graph)

11. Unsaturated, about 240 g of sugar are soluble per 100 g H_2O at 40°C.

12. Saturated; interpolation from graph shows that only about 130 g of $NaNO_3$ are soluble in 100 g H_2O at 60°C.

13. (a) $M = 4.0\ \text{mol}/8.0\ \text{L} = \underline{0.50\ M}$
    (b) $M = \frac{180\ \text{g}}{(180\ \text{g/mol})(0.500\ \text{L})} = \underline{2.00\ M}$

14. (a) $M = 1.0\ \text{mol}/5.0\ \text{L} = \underline{0.20\ M}$
    (b) $M = \frac{3.52\ \text{g}}{(176\ \text{g/mol})(1.00\ \text{L})} = \underline{0.0200\ M}$

15. (a) Basic; $(1.0 \times 10^{-14}\ M^2)/(1.0 \times 10^{-9}\ M) = 1.0 \times 10^{-5}\ M$; 9
    (b) Acidic; $(1.0 \times 10^{-14}\ M^2)/(1.0 \times 10^{0}\ M) = 1.0 \times 10^{-14}\ M$; 0

16. (a) Neutral; $(1.0 \times 10^{-14}\ M^2)/(1.0 \times 10^{-7}\ M) = 1.0 \times 10^{-7}\ M$; 7
    (b) Acidic; $(1.0 \times 10^{-14}\ M^2)/(1.0 \times 10^{-3}\ M) = 1.0 \times 10^{-11}\ M$; 3

17. $V_A = \frac{20.00\ \text{mL} \times 1 \times 1.00\ M}{3 \times 10.00\ \text{mL}} = \underline{0.667\ M}$

18. $V_A = \frac{30.00\ \text{mL} \times 1 \times 2.00\ M}{1 \times 15.00\ \text{mL}} = \underline{4.00\ M}$

19. Formula masses are 64.1 u, 28.0 u, and 39.9 u, respectively. The lowest FM means the fastest gas, and so the answer is CO.

20. Formula masses are 16.0 u, 17.0 u, and 28.0 u, respectively. The lowest FM means the fastest gas, and so the answer is CH_4

21. One-third as much ($p \propto 1/V$)

22. Double ($p \propto T$)

23. 24 times ($p \propto nT/V$)

24. 3 times greater ($p \propto n/V$)

25. $V = \frac{0.50\ \text{mol} \times 0.0821\ \text{L-atm/mol-K} \times 300\ \text{K}}{2.0\ \text{atm}} = \underline{6.2\ \text{L}}$

26. $T = \frac{4.00\ \text{atm} \times 10.0\ \text{L}}{2.00\ \text{mol} \times 0.0821\ \text{L-atm/mol-K}} = \underline{244\ \text{K}}$

ANSWERS TO RELEVANCE QUESTIONS

14.1. The lowest common denominator of 8 and 12 is 24, so get three 8-packs of buns and two 12-packs of wieners.

14.2. You are suprised by acid rain being reported with pH of 9.6 because that pH is above 7 and thus indicates a "base" rain.

14.3. Unless two holes are punched in the can lid, a partial vacuum will form as the juice is poured, and the atmospheric pressure will tend to hold the juice in the can. The second hole allows air to flow into the can to equalize the pressure outside and in.

ANSWERS TO STUDY GUIDE QUIZ

Multiple-Choice Questions

1. b	2. c	3. b	4. c	5. b
6. c	7. d	8. c	9. b	10. b

Short-Answer Questions

1. When a diver is deep under water, the pressure increases, and nitrogen is dissolved in his bloodstream to a much higher degree than would be the case when he was at the surface. Returning rapidly to the surface releases the dissolved nitrogen into the bloodstream in such a way that bubbles that can block the passage of blood through small blood vessels and cut off circulation may form. A more gradual reduction in pressure will not produce this adverse effect.

2. FU stands for formula unit. A formula unit can be any basic entity under study, such as an atom, a molecule, or an electron.

   MM represents molar mass, that is, the number of grams of an element or compound that is numerically equal to the formula mass of that material.

   M is the molarity, or the number of moles of solute per liter of solution.

   $[H^+]$ stands for the concentration of H^+ ions in moles per liter that is present in any solution. This number is related to the pH of the solution and determines if the solution is acidic or basic.

3. First find the MM for laughing gas, 2×14.0 g (for two nitrogen atoms) + 16.0 (for one oxygen atom) = 44.0 g. The number of moles of gas will then be 88.0 g/44.0 g = 2.00 mol.

4. The number of moles can be found by dividing the number of entities (molecules in this case) by Avogardo's number. # Moles = $(18.1 \times 10^{23})/(6.02 \times 10^{23}) = 3.00$ mol

5. Solubility is the amount of solute that will dissolve in a specified volume or mass of solvent, at a given temperature, to produce a saturated solution.

   The solvent is the substance present in greatest amount in the solution, and the solute is the substance dissolved in this solvent.

6. First find the MM for NaCl = 23.0 g for Na + 35.5 g for Cl = 58.5 g

   The molarity is then M = 29.3 g/[(58.5 g/mol) (0.500 L)] = 1.00 M

7. $P \propto nT/V$, so $P \propto (1)(2)/(4) = 1/2$; the new pressure is one-half the old pressure.

8. $P = nRT/V$ = (5.00 mol) (0.0821 L-atm/mol-K) (27° + 273°) / 60.0 L = 2.05 atm

9. $V_a = V_b M_b n_b / M_a n_a$ = (12.00 mL) (0.500 M) (2) / [(0.600 M)(1)] = 20.0 mL

10. pH = – log $[H^+]$ = –(–4) = 4

    $[OH^-] = (1.0 \times 10^{-14}\ M^2) / [H^+] = (1.0 \times 10^{-14}\ M^2) / (1.0 \times 10^{-4}\ M) = 1.0 \times 10^{-10}$ M

Chapter 15

Organic Chemistry

This chapter deals with the division of chemistry that probably most affects our daily lives. Organic compounds are all around us, and, in fact, basically *are* us. Some instructors may not feel comfortable teaching organic chemistry, but every effort has been made to make the presentation of the chapter material clear, consistent, coherent, and understandable.

Biochemistry is not emphasized because that material is covered in biology courses. Ester and amide formation reactions are discussed because they illustrate organic synthesis and are reactions that are important in preparing drugs, polymers, fats, and proteins.

For success in handling this chapter, the student must learn at once the basic bonding rules in Table 15.1 (these were also covered in Chapter 12) and the names and molecular formulas of the first eight alkanes (Table 15.2). In order for the student to understand simple isomers, the instructor must continually emphasize the tetrahedral geometry of the four single bonds to carbon, and that there is no difference between drawing a bond to carbon "up," "down," or "sideways" on the two-dimensional blackboard.

If time is short, the instructor could stop at any point in the chapter, but each section builds on the preceding sections.

DEMONSTRATIONS

The use of ball-and-stick models to show bonding and simple substitution, addition, and ester-and-amide formation reactions is recommended. Also, isomers are most easily discussed with the aid of such models.

A simple demonstration reaction is the addition of Br_2 to an alkene. Use bromine water and a liquid alkene such as cyclohexene. The reaction takes place at room temperature, and the characteristic color of bromine disappears as the reaction takes place. Take care: Cyclohexene is flammable, and bromine water irritates the eyes. For additional ideas on demonstrations, refer to the publications listed in the Teaching Aids section at the end of this *Guide*.

The film or video *Hydrocarbons and Their Structures*, available from Coronet, fits in well with this chapter. For additional ideas on appropriate audio-visual material, refer to the catalogs of the audio-visual suppliers listed in the Teaching Aids Section at the end of this *Guide*.

ANSWERS TO REVIEW QUESTIONS

1. Organic chemistry studies the compounds that contain carbon. Biochemistry studies the chemical compounds and reactions that occur in living cells.
2. d

3. Carbon (4 bonds), hydrogen (1 bond), oxygen (2 bonds), sulfur (2 bonds), nitrogen (3 bonds), a halogen (1 bond).
4. a
5. b
6. An aromatic hydrocarbon contains one or more benzene rings.
7. Alkanes, C_nH_{2n+2}, CH_3CH_3, etc.

   Cycloalkanes, C_nH_{2n}, ▽ , etc

   Alkenes, C_nH_{2n},, $CH_2 = CH_2$, etc.

   Alkynes, C_nH_{2n-2}, $HC \equiv CH$, etc.
8. Methane, CH_4 (the principal component of natural gas)

   Ethane, C_2H_6

   Propane, C_3H_8

   Butane, C_4H_{10}

   Pentane, C_5H_{12}

   Hexane, C_6H_{14}

   Heptane, C_7H_{16}

   Octane, C_8H_{18}
9. The bonds point toward the corners of a regular tetrahedron. They form angles of 109.5° to one another.
10. Structural isomers.
11. The end of the name, to determine the number of carbon atoms in the longest continuous chain (the "backbone").
12. Methane, CH_4,

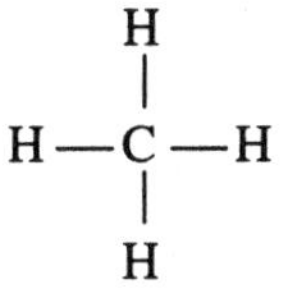

$$\begin{array}{ccccc} & & H & & \\ & & | & & \\ H & — & C & — & H \\ & & | & & \\ & & H & & \end{array}$$

Methyl group, CH_{3-},

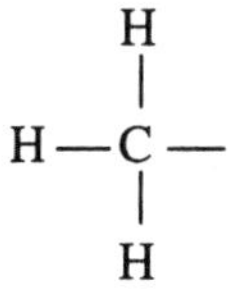

$$\begin{array}{cccc} & & H & \\ & & | & \\ H & — & C & — \\ & & | & \\ & & H & \end{array}$$

Ethane, CH_3CH_3,

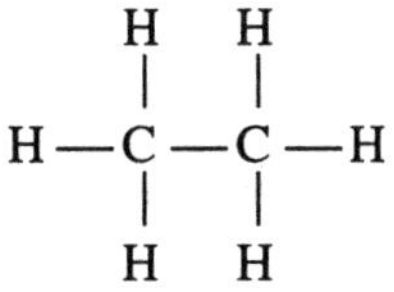

$$\begin{array}{ccccccc} & & H & & H & & \\ & & | & & | & & \\ H & — & C & — & C & — & H \\ & & | & & | & & \\ & & H & & H & & \end{array}$$

Ethyl group, CH_3CH_{2-},

$$\begin{array}{cccccc} & & H & & H & \\ & & | & & | & \\ H & — & C & — & C & — \\ & & | & & | & \\ & & H & & H & \end{array}$$

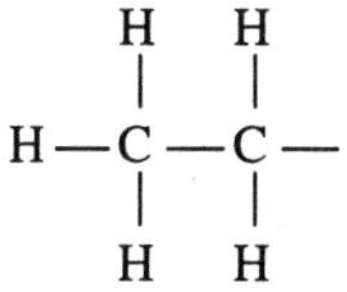

13. *R* stands for an alkyl group.
14. Cyclobutane, C_4H_{10},

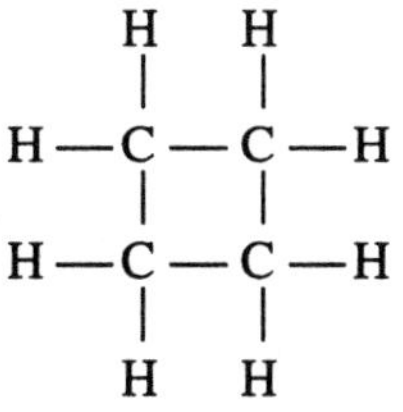

$$\begin{array}{ccccccc} & & H & & H & & \\ & & | & & | & & \\ H & — & C & — & C & — & H \\ & & | & & | & & \\ H & — & C & — & C & — & H \\ & & | & & | & & \\ & & H & & H & & \end{array}$$

15. Ethylene, $H_2C{=}CH_2$.

    Acetylene, $HC{\equiv}CH$.

16. Saturated hydrocarbons are those whose hydrogen content is at a maximum. Unsaturated hydrocarbons have double or triple bonds, which allow the addition of more hydrogen atoms.

17. An alkene can add a molecule of H_2 or halogen to the double bond. An alkyne can add two molecules of H_2 or halogen to the triple bond. Alkanes cannot undergo addition reactions because to do so would make five bonds to a carbon atom, which can form only four.

18. d

19. Benzene is a liquid obtained from petroleum. Its molecular formula is C_6H_6, and its preferred structural formula is [benzene ring]

20. c

21. d

22. d

23. R–X

24. Chlorofluorocarbon. Used in air conditioners, refrigerators, and heat pumps. Upon reaching the stratosphere, CFCs provide chlorine atoms that can destroy ozone molecules that protect life by absorbing UV radiation.

25. Alcohols have the general formula R–OH, contain one or more hydroxyl groups (—OH), and have names ending in *-ol.*

26. Amines have the general formula $R{-}NH_2$, contain one or more amino groups ($-NH_2$), and have an unpleasant odor.

27. The general formula for a carboxylic acid is RCOOH. It contains a carboxyl group (—COOH). CH_3COOH is the condensed structural formula for acetic acid, which is found in vinegar.

28. The general formula for an ester is RCOOR′. Esters have pleasant odors. Methyl salicylate is found in wintergreen mints.

29. Carboxylic acids and alcohols react to form esters. The other product is water. The general equation for the reaction is

$$\underset{\text{Carboxylic acid}}{R{-}\overset{\overset{\displaystyle O}{\|}}{C}{-}O{-}H} + \underset{\text{Alcohol}}{H{-}O{-}R'} \xrightarrow{H_2SO_4} \underset{\text{Water}}{H_2O} + \underset{\text{Ester}}{R{-}\overset{\overset{\displaystyle O}{\|}}{C}{-}O{-}R'}$$

30. Carboxylic acids and amines react to form amides. The other product is water. The general equation for the reaction is

$$\underset{\text{Carboxylic acid}}{R{-}\overset{\overset{\displaystyle O}{\|}}{C}{-}O{-}H} + \underset{\text{Amine}}{H{-}\overset{\overset{\displaystyle H}{|}}{N}{-}R'} \longrightarrow \underset{\text{Water}}{H_2O} + \underset{\text{Amide}}{R{-}\overset{\overset{\displaystyle O}{\|}}{C}{-}\overset{\overset{\displaystyle H}{|}}{N}{-}R'}$$

31. (a) Carbohydrates contain abundant hydroxyl groups.
    (b) Fats and oils are triesters of glycerol.
    (c) Proteins are polyamides.

32. Glucose and fructose combine to form sucrose. Glucose. Herbivores have bacteria in their digestive tracts that have the enzymes necessary to break the glucose linkages in cellulose.

32. Glucose and fructose combine to form sucrose. Glucose. Herbivores have bacteria in their digestive tracts that have the enzymes necessary to break the glucose linkages in cellulose.

33. Amino acids. The simplest is glycine,

$$H_2N-\underset{\underset{H}{|}}{\overset{\overset{H}{|}}{C}}-\overset{\overset{O}{\|}}{C}-OH$$

34. Both fats and oils are triesters of fatty acids and glycerol. In fats, the fatty acids are saturated; in oils, they are unsaturated (contain carbon-carbon double bonds).An oil is converted to a fat by adding hydrogen to the double bonds, a process called hydrogenation.

35. Soap is composed of the sodium salts of the fatty acids produced when fats are treated with sodium hydroxide.

    Both soaps and synthetic detergents have a long, nonpolar hydrocarbon part attached to a polar group. The polar group in soap is $-COO^- Na^+$; the polar group in detergents is $-OSO_3^- Na^+$.

36. b

37. a

38. c

39. *Addition polymers* are formed when alkene monomers add to one another.

    Condensation polymers are constructed from molecules that have two or more reactive groups. Each molecule attaches to two others by ester or amide linkages.

40. Dacron is a polyester, and nylon is a polyamide. Common addition polymers include polyethylene, Teflon, Styrofoam, and polyvinyl chloride

ANSWERS TO CRITICAL THINKING QUESTIONS

1. Celery contains an abundance of cellulose, which cannot be digested by humans.
2. The amines (vitamins were originally called vit*amines*).
3. 1-butyne would already have four bonds (one triple and one single) to the second carbon, and so it would be impossible to attach a methyl group to that carbon.
4. Peel the onion under the surface of water in a bowl.
5. 2-bromo-2-chloro-1,1,1-trifluoroethane is the name that puts the three substituents in alphabetical order and gives the lowest set of position numbers.
6. The element silicon, Si, is in the same group and is a nonmetal like carbon. However, its bonding capabilities do not come close to those of carbon.

ANSWERS TO EXERCISES

1. (a) is valid.
   (b) is bogus because Cl should have one bond, not two.
2. Formula b is bogus, because the O has three bonds when it should have only two and the C has three bonds when it should have four.

3. (a) Alkene.
   (b) Cycloalkane.
   (c) Alkyne.
   (d) Alkane.
   (e) Aromatic.

4. (a) Alkane.
   (b) Aromatic.
   (c) Cycloalkane.
   (d) Alkene.
   (e) Alkyne.

5. (a) Alkane.
   (b) Alkyne.
   (c) Cycloalkane.
   (d) Alkene.
   (e) Aromatic.

6. (a) Alkene.
   (b) Cycloalkane.
   (c) Alkane.
   (d) Aromatic.
   (e) Alkyne.

7. (a) Same compound.
   (b) Neither.
   (c) Isomers.

8. (a) Neither.
   (b) Same compound.
   (c) Isomers.

9. (a) 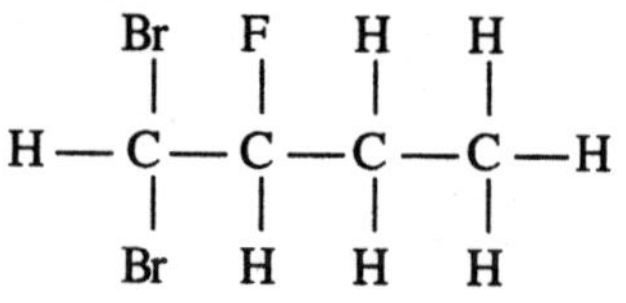 $H-CBr_2-CFH-CH_2-CH_2-H$ (b) 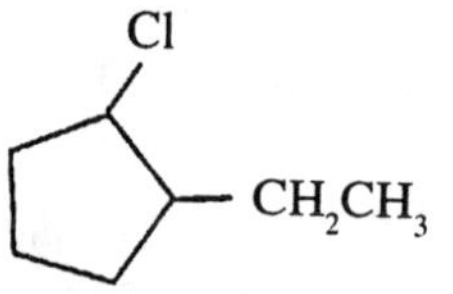 Cl, CH_2CH_3 (c) 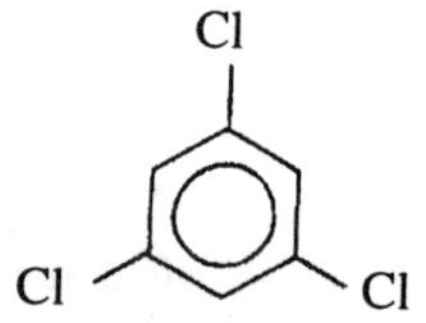 Cl, Cl, Cl

10. (a) 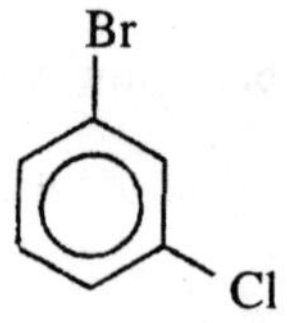 Br, Cl (b) 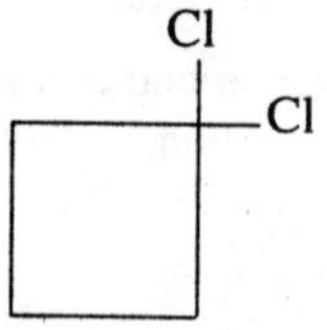 Cl, Cl (c) 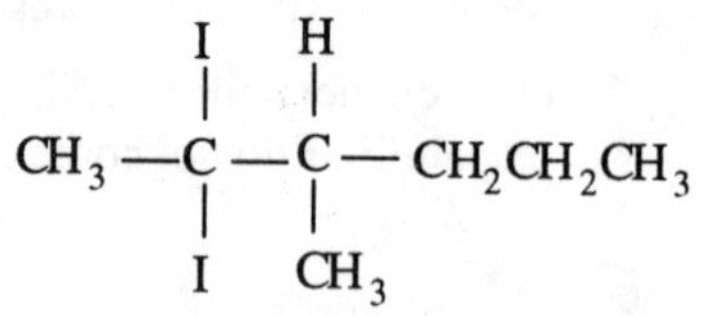 $CH_3-CI_2-CH(CH_3)-CH_2CH_2CH_3$

11. 1-pentene

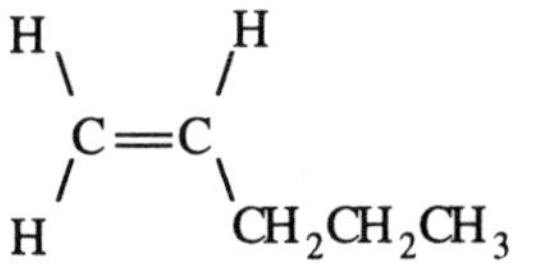

$$\begin{matrix} H & & H \\ & C{=}C & \\ H & & CH_2CH_2CH_3 \end{matrix}$$

and 2-pentene

$$\begin{matrix} CH_3 & & CH_2CH_3 \\ & C{=}C & \\ H & & H \end{matrix}$$

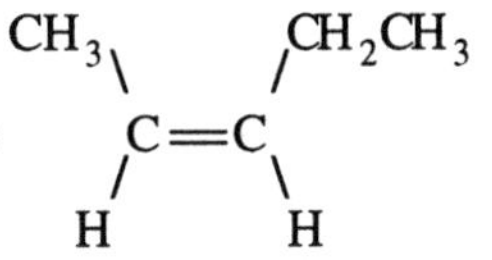

(3-Pentene would be the same as 2-pentene, and 4-pentene the same as 1-pentene.)

12. three isomers: 1-hexene, $CH_2{=}CHCH_2CH_2CH_2CH_3$
2-hexene, $CH_3CH{=}CHCH_2CH_2CH_3$
3-hexene, $CH_3CH_2CH{=}CHCH_2CH_3$

13. (a)

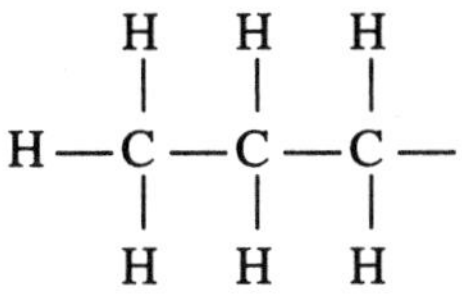

$$\begin{matrix} & & H & & H & & H & & \\ & & | & & | & & | & & \\ H & — & C & — & C & — & C & — & H \\ & & | & & | & & | & & \\ & & H & & H & & H & & \end{matrix}$$

; propane

(b)

$$\begin{matrix} & & Cl & & Cl & & \\ & & | & & | & & \\ CH_3 & — & C & — & C & — & CH_3 \\ & & | & & | & & \\ & & Cl & & Cl & & \end{matrix}$$

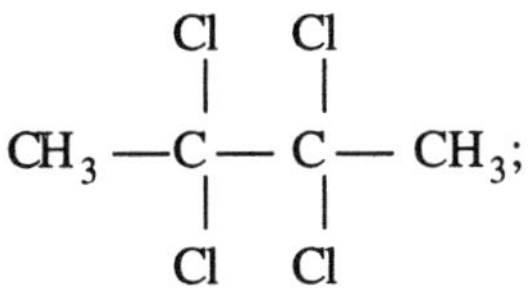

; 2,2,3,3-tetrachlorobutane

14. (a) 1,2-Dibromopropane; $CH_2BrCHBrCH_3$
(b) Butane; $CH_3CH_2CH_2CH_3$

15. (structures: benzene rings with CH_3 substituents) and 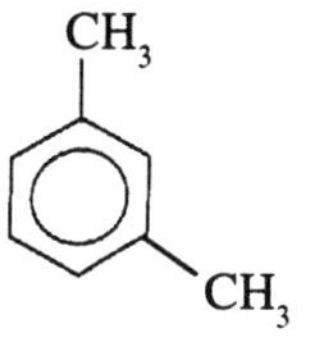 … and …

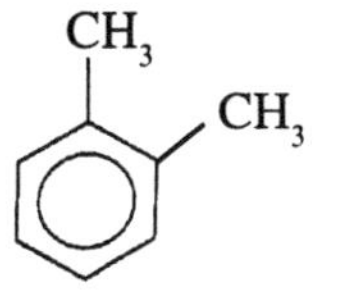

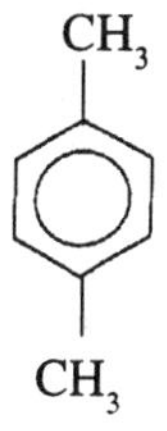

1,2-dimethylbenzene   1,3-dimethylbenzene   1,4-dimethylbenzene

16. (structures: benzene rings with CH_2CH_3 substituents)

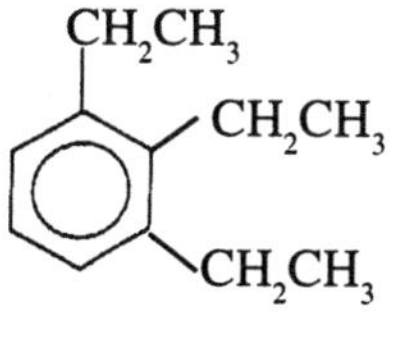

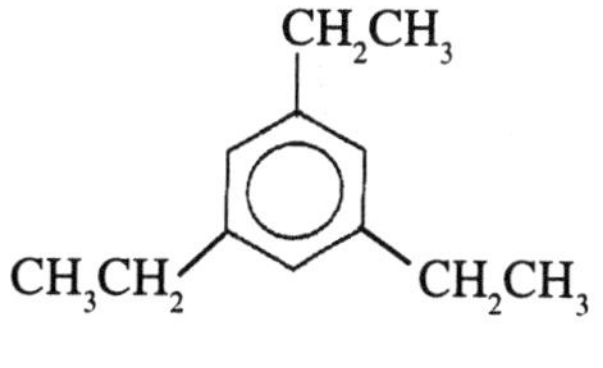

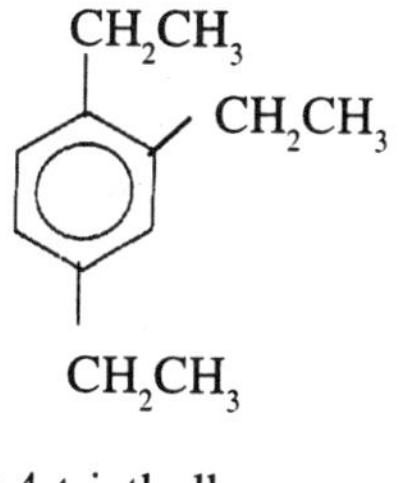

1,2,3-triethylbenzene   1,3,5-triethylbenzene   1,2,4-triethylbenzene

17. (a) Amine
    (b) Amide
    (c) Ester
    (d) Carboxylic acid
    (e) Alkyl halide
    (f) Alcohol

18. (a) Carboxylic acid
    (b) Ester
    (c) Alcohol
    (d) Alkyl halide
    (e) Amine
    (f) Amide

19. (a) $CH_3CH_2—OH$ and $CH_3—O—CH_3$
    (b) $CH_3CH_2CH_2NH_2$, $CH_3CH(NH_2)CH_3$, $CH_3CH_2NHCH_3$, $(CH_3)_3N$

20. (a)

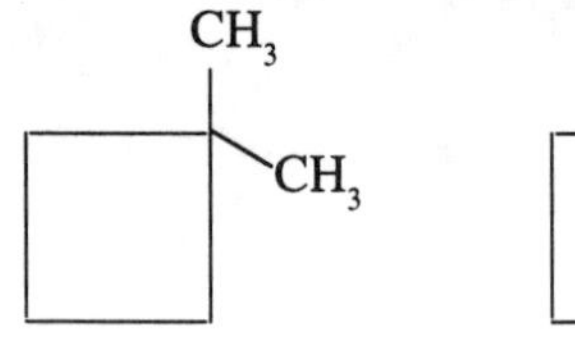

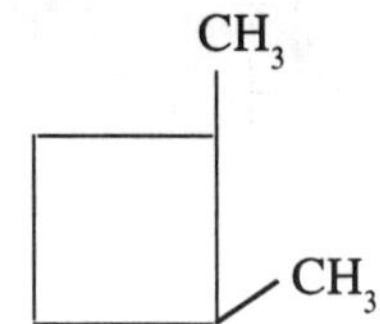

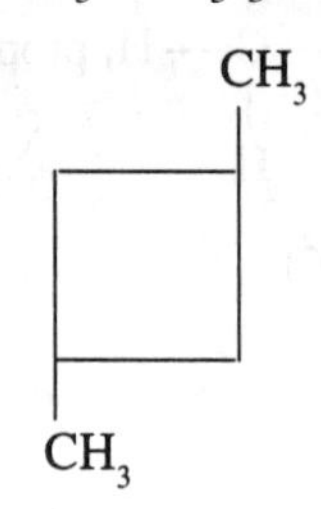

(b) $CH_3CH_2CH_2CH_2CH_3$

$CH_3CHCH_2CH_3$ (with CH_3 on the second carbon)

$CH_3—C(CH_3)_2—CH_3$

21. (a) $CH_3CH_2C(=O)—O—CH_2CH_3 + H_2O$

(b) $C_6H_5—CH_2C(=O)—N(H)—CH + H_2O$

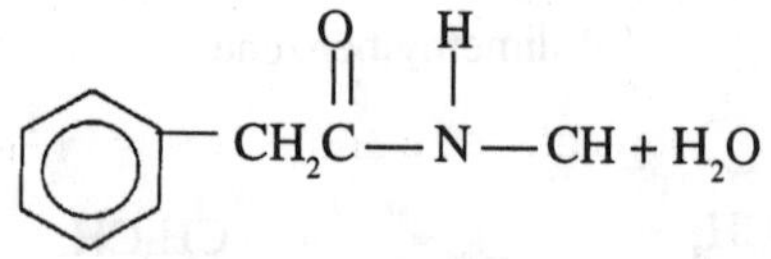

22. (a) $CH_3CH(CH_3)C(=O)—O—CH_2—C_6H_5 + H_2O$

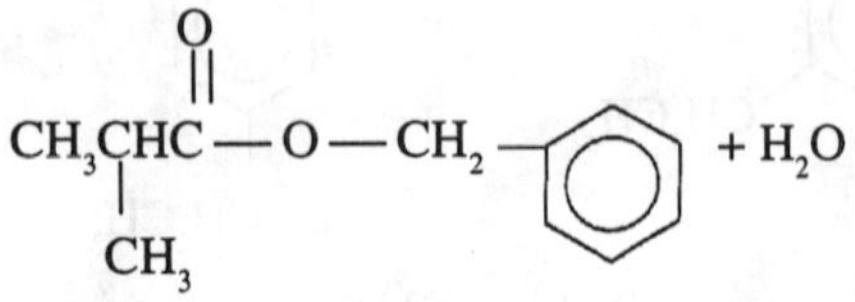

(b) $CH_2(F)C(=O)—O—N(H)—CH_2CH_2—C_5H_9 + H_2O$

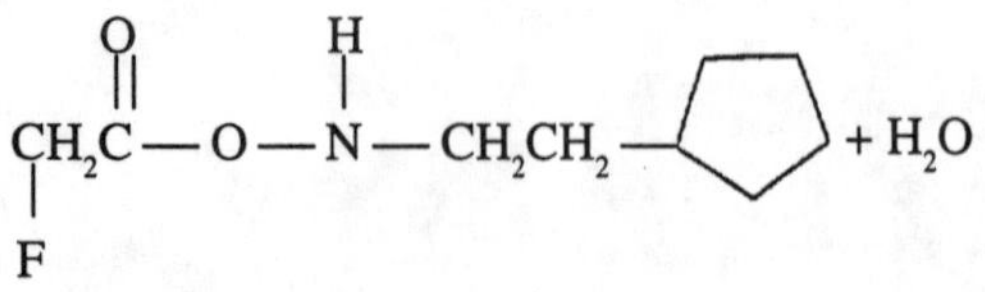

23. $\mathrm{C_6H_5(H)C{=}C(H)(H)} \rightarrow \left[-\mathrm{C(H)(C_6H_5)-C(H)(H)}-\right]_n$

24. $\mathrm{(H)(H)C{=}C(H)(CN)} \rightarrow \left[-\mathrm{C(H)(H)-C(H)(CN)}-\right]_n$

25. $\left[-\mathrm{HN-C_6H_4-NH-C({=}O)-C_6H_4-C({=}O)}-\right]_n$

26. $\left[-\mathrm{O-CH(CH_3)C({=}O)-O-CH(CH_3)C({=}O)}-\right]_n$

ANSWERS TO RELEVANCE QUESTIONS

15.1. The soot and other hydrocarbons such as benzene and other aromatic hydrocarbons could cause respiratory diseases, lung disease, or cancer. Also, the soot signals incomplete combustion, and so a lot of invisible carbon monoxide is also present.

15.2. Decaying flesh contains a lot of *amines*, such as the cadaverine and putresine mentioned in the chapter.

ANSWERS TO STUDY GUIDE QUIZ

Multiple-Choice Questions

1. b	2. c	3. a	4. a	5. b
6. b	7. a	8. d	9. b	10. c

Short-Answer Questions

1. Hydrocarbons can either contain benzene rings, in which case they are called *aromatic hydrocarbons*, or not contain benzene rings, in which case they are referred to as *aliphatic hydrocarbons*.
2. The alkenes have only single carbon-to-carbon bonds in their structures, the alkenes have one double carbon-to-carbon bond in their structures, and the alkynes have one triple carbon-to-carbon bond in their structures.
3. The alkyl halides have a halogen atom such as chlorine, fluorine, or iodine attached to the hydrocarbon chain. The general formula for these molecules is R — X, where R represents the hydrocarbon portion of the molecule and X represents the halogen portion.
4. (a) Alcohols have the general formula R — OH.
   (b) Carboxylic acids have the general formula RCOOH.
   (c) Alkanes have the general formula C_nH_{2n+2}.
   (d) Alkynes have the general formula, C_nH_{2n-2}.
5. Methane, propane, butane, and octane belong to the *alkane* series of hydrocarbons.
6. A benzene ring is a collection of six carbon atoms that form a circular (ringlike) structure that is the fundamental building block for many hydrocarbons and derivatives of hydrocarbons. Advanced bonding theory indicates that the bonding between these six carbon atoms is not structured as three single bonds and three double bonds, as was once thought, but is characterized by the sharing of six electrons among all of the carbon atoms in the ring.
7. *Plastics* are a group of synthetic polymers (long, repeating chains of smaller molecules) that can be molded and hardened to produce clothing, shoes, buildings, car parts, sports equipment, toys, etc.
8. A *detergent* is a soap substitute made up of a long hydrocarbon chain that is nonpolar, combined with a polar group such as sodium sulfate. Detergents are referred to as *synthetic* because there are no molecules in nature that have the same molecular structure.
9. *Structural isomers* are compounds that have the same molecular formulas but differ in structure. They can be formed when the same number and type of atoms can be assembled in more than one structural pattern without violating the octet rule of bonding.
10. *Proteins* are extremely long-chain polyamides formed by the enzyme-catalyzed polymerization of amino acids. Proteins perform important functions in living organisms and can be formed in living cells under the direction of nucleic acids or in biochemical laboratories.

Chapter 16

The Solar System

This chapter begins with a definition of astronomy and the universe, followed by some historical comments and information about the instruments modern technology has provided today's astronomers in their quest for knowledge of the universe in which we live. This is followed by an overview of the solar system, which sets the stage for comparing and contrasting the properties of the planets. Tables 16.1 and 16.2 are presented to summarize the properties of and comparisons between the terrestrial and Jovian planets. The information given is up to date. We have included the latest reported information concerning four newly discovered planetary systems beyond our solar system. An illustration is shown that compares the new planetary systems to the terrestrial planets.

The theory on the origin of the solar system is the best explanation we have, based on present knowledge. The authors present this theory in conjunction with the birth of stars as studied in Chapter 19. The instructor may find it helpful to assign Chapter 19 as a reading assignment when Chapter 16 is studied.

DEMONSTRATIONS

Model solar system that shows motion of planets.

Foucault pendulum. Use square frame (2 ft on each side) and clamp to turntable, with pendulum attached at top center.

Large earth sphere, to illustrate rotation and revolution.

Large sphere with slate surface to illustrate Eratosthenes' calculation.

Class demonstration of parallax (finger of each hand held together, then one held close to the face and the other at arm's length).

ANSWERS TO REVIEW QUESTIONS

1. d
2. b
3. c
4. The Sun, nine planets, 63 known moons, thousands of asteroids, vast numbers of comets and meteoroids, a solar wind, gases, and interplanetary dust particles.
5. A geocentric model is Earth-centered; a heliocentric model is Sun-centered.

6. The law of conservation of angular momentum is the physical explanation of Kepler's second law. As an object approaches the Sun, it must speed up to conserve angular momentum. A careful analysis shows that this causes equal areas to be swept out in equal times.
7. Prograde motion is orbital or spin motion in the forward or direct direction; in the solar system, this is west-to-east (eastward), or counterclockwise as viewed from above the North Pole. Retrograde motion is backward motion; in the solar system, this is an apparent east-to-west (westward) motion of a planet with respect to the stars.
8. The Titius-Bode law does not represent a physical property of the solar system.
9. The orbital speed of a planet increases as the distance from the Sun decreases (conservation of angular momentum).
10. Sidereal period is the orbital or rotational period of an object with respect to the stars as observed from the Sun. Synodic period is the orbital or rotational period of an object as observed from Earth.
11. Conjunction.
12. b
13. c
14. An albedo of 0.33 means that 33 percent of incoming sunlight is reflected from Earth's surface.
15. Rotation is the motion of Earth around an axis; it is counterclockwise as viewed from above the North Pole. Revolution is the motion of Earth about the Sun; it is counterclockwise as viewed from above the North Pole.
16. Rotation—the motion of a Foucault pendulum. Revolution—the occurrence of parallax.
17. The ecliptic is the apparent path of the Sun on the celestial sphere.
18. A parsec is a measure of distance. One parsec is defined as the distance to a star when the star exhibits a parallax of one second of arc.
19. The parallax of a star cannot be seen with the unaided eye, because the stars are very far away compared to the distance between Earth and the Sun. The parallax angle is exceedingly small and so cannot be detected with the unaided eye.
20. Revolution.
21. One astronomical unit is the average distance between Earth and the Sun, 1.5×10^8 km.
22. d
23. c
24. Mercury, Venus, Earth, and Mars. They are called the terrestrial planets because they resemble Earth in general physical and chemical properties.
25. Mercury has an extremely thin atmosphere of sodium and potassium, believed to be ejected form the planet's surface by the solar wind.
26. Venus has a dense atmosphere whose composition is 98% carbon dioxide, which produces the "greenhouse effect."
27. The ancient channels on Mars indicate that fluid once flowed on the Martian surface.
28. Venus has an atmospheric pressure slightly less than 1% that of Earth.
29. They are inferior planets; their orbits are less than Earth's.
30. At full phase Venus is opposite the Sun (180° from Earth) and cannot be seen because of the brightness of the Sun. When Venus is closest to Earth, it is between Earth and the Sun, and the dark side is toward Earth. During both times given in the question, Venus is on the same meridian as the Sun.

31. d
32. a
33. Jupiter, Saturn, Uranus, Neptune.
34. All Jovian planets are believed to have rocky cores (composed mainly of iron, oxygen, and silicon), with a layer of ice (materials composed of carbon, nitrogen, and oxygen in combination with hydrogen) above the rocky core. All have outer layers of molecular hydrogen. Jupiter and Saturn have a layer of metallic hydrogen below the molecular hydrogen. See Fig. 16.23.
35. Roche limit.
36. Tidal force is a differential gravitational force that tends to deform or stretch a body.
37. Io is the only satellite in the solar system to have active volcanoes.
38. Titan is the only satellite in the solar system known to have a dense, hazy atmosphere.
39. Pluto is classified as a planet because it orbits the Sun. Pluto should not be classified as a planet because its orbit is highly irregular—it is noncircular and greatly tilted to the ecliptic plane. The orbit moves inside that of Neptune at times, and Pluto does not resemble the other outer planets.
40. Titan's surface temperature is much lower than Ganymede's because of its greater distance from the Sun. The lower temperature decreased molecular speeds, which decreased their rate of escape. Also, Titan outgassed more gas from its interior than the Jovian satellite.
41. They are similar in composition, except that Uranus and Neptune contain no metallic hydrogen. See Fig. 16.23 and the answer to Question 34.
42. The magnetic field of both planets is tipped greater than 55° to the rotational axis, and offset from the center of the planet.
43. Refer to Table 16.2 for several significant differences between the terrestrial and Jovian planets.
44. On the average, Pluto, but it is in a highly elliptical orbit and actually goes inside Neptune's orbit, so that Neptune can actually be farther from the Sun than Pluto. At present, Neptune is the most distant planet from the Sun. (See Fig. 16.6.)
45. Pluto.
46. Jupiter; Jupiter's Ganymede.
47. Phobos, Mars; Ganymede, Jupiter; Titan, Saturn; Triton, Neptune; Miranda, Uranus.
48. Mercury, Pluto.
49. Mercury, Venus, Mars, Saturn, and Jupiter.
50. Jupiter, 5 h.
51. Titan.
52. Jupiter, Saturn, Uranus, and Neptune.
53. e
54. e
55. An asteroid is an object orbiting the Sun that is smaller than a major planet. Meteoroids are small interplanetary objects in space before they encounter Earth.
56. Ceres. Ceres (a carbonaceous class asteroid) is slightly more than 940 km in diameter.
57. More than 2000 asteroids have been named and numbered. Most orbit the Sun between Mars and Jupiter.

58. (a) A small irregular object (typical diameter 5 to 20 km) composed mainly of dust and ice.
    (b) Close to the Sun, the surface of the comet vaporizes to form a gaseous head and a long plasma tail.
59. Most observed comets originated in the source region called the Oort Cloud. This is a spherical region around the Sun at about 50,000 AU.
60. The Oort Cloud is a spherical region around the Sun some 50,000 AU from the Sun. This is believed to be the source region for most observed comets.
61. Meteoroids are small interplanetary objects that have not entered Earth's atmosphere. Meteors are meteoroids that enter Earth's atmosphere and appear as "shooting stars." Meteorites are meteoroids that strike Earth's surface.
62. The observation of zodiacal light and the gegenshein.
63. 80,000 to 100,000 AU.
64. e
65. d
66. The primordial nebula is a large swirling volume of gas and dust located in space among the stars of the Milky Way. The solar nebula is the flattened rotating disk of gas and dust around the protosun, from which the planets formed. Some astronomers call the solar nebula the primitive solar system.
67. Dust particles act as condensation nuclei for the collection of atoms and molecules of interstellar gas. The dust particles speed up the condensation process to form planets.
68. The best evidence comes from meteorites, which are the least-changed fragments of the initial solar system.
69. The law of conservation of angular momentum demands that the momentum of the interstellar atoms and molecules be conserved; thus, the contracting matter must form a rotating mass.
70. 10^8 years.
71. c
72. b
73. The detection of a Doppler shift in the spectrum of a star.
74. The observation of a Doppler shift in the spectrum of the pulsar.
75. The Billion-channel Extraterrestrial Array to detect radio signals from outer space.
76. The entire water hole (1.42 to 1.67 GHz) will be covered.

ANSWERS TO CRITICAL THINKING QUESTIONS

1. The pendulum continues to swing in a north-south plane during the 24-h period. No deviation will be observed.
2. The outer planets are never between Earth and the Sun; therefore, they never pass through a new phase position.
3. The planets are too far apart to cast a shadow on one another.

4. The plane of Saturn's rings is tilted about 27° to the plane of Saturn's orbit. Thus, when Saturn is at positions P_1 and P_3 in the drawing, its rings will be observed edgewise. At position P_1, the northern side will be observed. At P_4, the southern side will be observed.

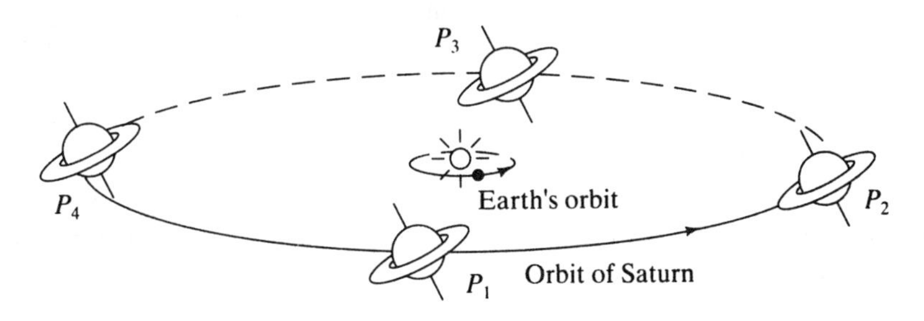

5. Phobos' revolving period (7 h, 39 min) is shorter than Mars' rotation period (24.6 h). Since Phobos revolves eastward, it will rise on Mars' western horizon.

ANSWERS TO EXERCISES

1. $T^2 = kR^3 = 1.00 \dfrac{y^2}{(AU)^3} \times \dfrac{2^3(AU)^3}{1} = 8\ y^2$

   $T = \sqrt{8\ y^2} = \underline{2.8\ y}$

2. $T^2 = kR^3 = 1.00 \dfrac{y^2}{(AU)^3} \times \dfrac{4^3(AU)^3}{1} = 64\ y^2$

   $T = \sqrt{64\ y^2} = \underline{8\ y}$

3. $T^2 = \left(\dfrac{4\pi^2}{G}\right) \dfrac{R^3}{m_{Sun}}$

   Let $K = \left(\dfrac{4\pi^2}{G}\right) = 1$ R remains at 1 AU

   Then $T^2 = 1 \times \dfrac{1}{m_{Sun}} = \dfrac{1\ y^2}{0.5 \text{ molar mass}}$

   $T = \sqrt{\dfrac{1\ y^2}{0.5}} = \underline{1.4\ y}$

4. $T = \sqrt{\dfrac{y^2}{4}} = \underline{0.5\ y}$

5. $\dfrac{T^2}{R^3} = k = \text{constant};\ \therefore\ T^2 = k \times R^3.$

   Therefore, the smaller R is, the smaller T is.

6. $\dfrac{T^2}{R^3} = k$ or $\dfrac{R^2}{T^2} = \dfrac{1}{k \times R}, \dfrac{(2\pi R)^2}{T^2} = \dfrac{(2\pi)^2}{k \times R}$

   Therefore, $v^2 = \dfrac{(2\pi)^2}{k \times R}$, and the smaller R is, the bigger v is.

7. Titius-Bode law: 0.4, 0.7, 1.0, 1.6, 2.8 (asteroids) astronomical units.

   Actual: 0.39, 0.72, 1.0, 1.52, 2.77 (asteroids) astronomical units.

8. (a) Titius-Bode law: 5.2, 10.0, 19.6, 38.8, astronomical units.
   Actual: 5.2, 9.54, 19.19, 30.07 astronomical units.
   (b) Titius-Bode law: 77.2 astronomical units.
   Actual: 39.46 astronomical units.

9.

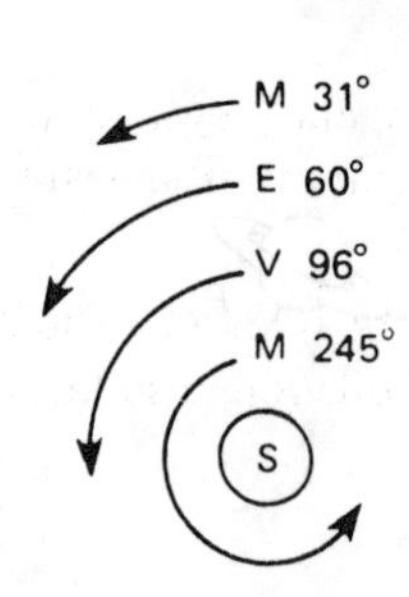

10.

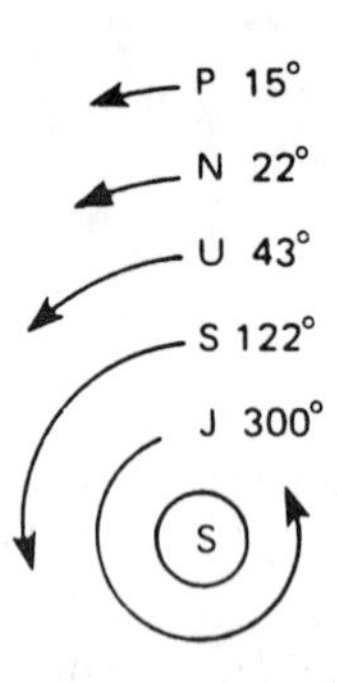

11. Mercury, Venus, Earth, Mars, Jupiter, Saturn, Uranus, Neptune, Pluto, although right now and until 1999, Pluto is actually closer to the Sun than Neptune.

12. (a) Sun to Earth: 93 million miles.
    (b) Mars to Jupiter: 341 million miles.
    (c) Jupiter to Saturn: 403 million miles.
    (d) Saturn to Uranus: 897 million miles.

13. Jupiter, Saturn, Uranus, Neptune, Earth, Venus, Mars, Mercury, Pluto.

14. Earth, Mercury, Venus, Mars, Neptune, Jupiter, Uranus, Pluto, Saturn.

ANSWERS TO RELEVANCE QUESTIONS

16.1 Answers will range from no impact to a strong interest in learning more about the solar system. Interested students may ask the instructor for additional reading material.

16.2 Rotation period—awake and sleep time; work and recreational time. Revolving period—reference for measuring age; seasonal changes on body.

16.3 (a) Oxygen, water, and organic molecules; sunlight, heat, water, and the correct temperature range.
     (b) Earth has the right orbit and distance from the Sun to provide the correct temperature range, and the planet's gravity is sufficient to retain an atmosphere and liquid water.

16.4 Answers will vary from "I don't know" to a catastrophic event. Planet Earth accumulates about 100 k of extraterrestrial material every day as a result of a constant temperature by interplanetary accumulates debris. In 1908 a 50-m-diameter rocky asteroid exploded in the atmosphere above a desolate region of central Siberia, devastating more than 2000 km^2 of forest.

16.5 A meteorite.

16.6 The most generally accepted star-formation theories suggest that planetary system are formed around young stars condensing from interstellar gas and dust, but special conditions are believed necessary to form solar systems similar to ours.

16.7 Most students are most likely to give a positive answer based on the thought, why should we be unique? For a statistical answer, refer the student to the Green Bank Equation, also known as the Drake Equation.

ANSWERS TO STUDY GUIDE QUIZ

Multiple-Choice Questions

1. b	2. d	3. a	4. d	5. d
6. d	7. c	8. d	9. c	10. b

Short-Answer Questions

1. A long pendulum suspended in a tall building changes its plane of swing throughout the day. In actuality, the plane of the pendulum's swing is always the same, but Earth rotates under it, providing the apparent change in direction of 15° every hour at the North or South Pole. After one day, the plane of swing has returned to its original orientation with respect to the building, showing that Earth has made one complete rotation in 24 h. This type of pendulum is called a Foucault pendulum. (See Section 16.2 for a more detailed explanation of this experiment.)
2. The fraction of solar radiation reflected by Earth back into space is referred to as Earth's albedo. If more light were reflected, say because of increased cloud cover or volcanic dust in the atmosphere, less energy would be absorbed by Earth's surface and the overall temperature of Earth's atmosphere would drop, causing a cooler climate. If less energy were reflected back into space, the reverse would be true.
3. Kepler's second law deals with the speed at which planets move as they revolve around the Sun. It is often referred to as the law of equal areas because of an imaginary line (radial vector) joining any planet to the Sun sweeps out equal areas in equal time periods. This requires that a planet travel faster along its orbit when it is closer to the Sun and slower when it is farther away.
4. Mercury is the closest planet to the Sun, followed by Venus, Earth, and Mars. These four planets make up the group known as the terrestrial planets.
5. A planet that is in opposition is located on exactly the opposite side of Earth from the Sun. Since the planets in our solar system orbit the Sun in very nearly the same plane, this happens regularly. Only the outer planets can ever be in opposition to the Sun as viewed from Earth.
6. Earth's atmosphere is composed primarily of nitrogen (about 78%) and oxygen (nearly 21%), with only small amounts of argon, carbon dioxide, and other gases. The water vapor content of Earth's atmosphere varies greatly with temperature, but never exceeds about 4%. Venus, on the other hand, is covered by a denser atmosphere composed of 96% carbon dioxide, around 4% nitrogen, and traces of argon, oxygen, and water vapor. The atmospheric pressure on Venus is about 90 times that of Earth, and the average surface temperature on Venus is much higher, around 750 K.

7. Asteroids are small, irregular, rocky or metallic objects that orbit about 2.8 AU from the Sun, most of them between the orbits of Mars and Jupiter. This cosmic debris is believed to be material that did not form into a planet the way most of the material surrounding the Sun did as our solar system formed.
8. When a space probe is sent to Mercury or Venus, the probe must lose energy. Since the launch platform (Earth) is moving at nearly 30 km/s in its orbit around the Sun, we can achieve this loss of energy by launching the probe in the direction opposite to the direction of Earth's orbit around the Sun.
9. Ring systems seem to be the rule, rather than the exception for the outer planets. The most obvious ring system belongs to Saturn and can be seen from Earth with any good-quality telescope on a clear night when Saturn is in the proper position in the sky. The rings are composed of ice and ice-coated rock and dust.
10. It is believed that our solar system began as a large swirling cloud of cold gases and dust in the form of a slowly rotating solar nebula. A process of condensation and accretion began, perhaps triggered by the shock wave of a nearby supernova, causing a collection of particles by mutual gravitational attraction. This process was slow at first, but speeded up as the central mass became larger. The rotational speed of the cloud flattened it into the equatorial plane, and the increased rotation rate, caused by conservation of angular momentum of the shrinking cloud, set up shearing forces. These shearing forces, aided by variations in density in the cloud itself, led to the formation of protoplanets. Over time, these protoplanets continued to condense and sweep up surrounding gases and dust to form the planets that orbit the Sun today.

Chapter 17

Place and Time

This chapter is very important and necessary for the prospective elementary school teacher. The authors have found that this chapter is of strong interest to the majority of students taking the course. Everyone is interested in knowing his or her position in space and time. The position of any object in space or time requires a frame of reference, and this chapter deals with the basic ideas of reference frames. It also explains the existence of seasons on our planet. Many students have an interest in determining the declination of the Sun for any day of the year. The instructor should have students make a simple astrolabe or sextant, which can be used to find the altitude of a star. Students will also gain a better understanding of standard time through the use of this instrument, because of declination of the Sun must be taken at 12 noon local time.

DEMONSTRATIONS

With colored chalk, draw the equator, small circles parallel to the equator, the prime meridian, and one or two other meridians on a 16-in. diameter slate sphere. Show the sphere axis tilted 23-1/2° down from the vertical. Place a light source representing the Sun at the center of the lecture disk, then move the slate sphere around the light, illustrating how the declination of the Sun changes as Earth revolves around the Sun. A model solar system is very good for showing this, but the system is too small for a large class to see it clearly.

Demonstrate the difference between the solar day and the sidereal day by having one of your students sit on a stool in front of the class to play the part of the Sun. You are Earth, and the other students are stars. Stand in front of the student on the stool facing the class and explain to the others the part they are playing. Then revolve about the student on the stool, keeping your back to the student at all times. After one revolution has been completed, ask the person on the stool how many rotations you made. The answer will be "None," because your back was in view at all times. Next ask the class (playing stars) how many rotations you made during one revolution. They will answer "One." Repeat the demonstrations but make one rotation with respect to the student on the stool. This will be two rotations with respect to the class. They will get the point that there is always one more rotation to anyone outside of the rotating and revolving system.

Demonstrate how the solar day varies in length by having the lamp on the lecture desk play the part of the Sun; you play the part of Earth. With a meter stick in your hand (this is more effective than just the extended arm), stand at the side of the desk facing the lamp (Sun). Hold the meter stick out toward the lamp so that, sighting over the meter stick, you observe the Sun. Next rotate counter-clockwise and revolve counter-clockwise (take a couple of steps) around the lecture desk and lamp. Rotate 360° with respect to your original position. Note that you cannot observe the sun by sighting across the meter stick but must rotate slightly more than 360° to see the Sun on your meridian. Repeat the demonstration. This time take three steps in revolving. Have the students note that you must rotate even more than last time because you have to move around farther in orbit. Thus, the length of a solar day depends on the orbital velocity of Earth. The greater the orbital velocity, the longer the solar day, provided that the rotation time remains constant.

ANSWERS TO REVIEW QUESTIONS

1. d
2. d
3. Every coordinate system must have a number line that indicates both a point of origin and unit length along the line.
4. Temperature scales, profit and loss, plus and minus declination.
5. 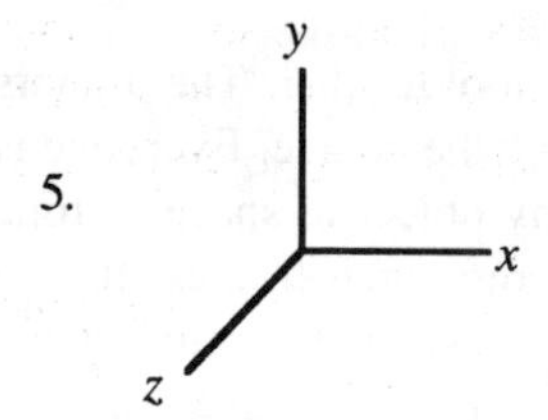

6. 90°
7. Answer is relative to student's location and reference point.
8. e
9. e
10. Latitude: 0° to 90°N and 0° to 90°S. Longitude: 0° to 180°E and 0° to 180°W.
11. Yes, because one is traveling in a circle that is parallel to the equator. Parallels are small circles.
12. No, because meridians are half-circles. They begin and end at the poles.
13. Latitude: At the equator. Longitude: At the prime meridian.
14. A meridian.
15. All meridians meet at the North Pole. The "longitude" of the North Pole has no meaning.
16. Sunrise about March 21; sunset about September 22.
17. All meridians are one-half of a great circle. They begin at one pole of Earth and end at the other.
18. All parallels are small circles except the equator, which is a great circle.
19. d
20. d
21. The rotation of Earth about an internal axis.
22. There are 24 time zones, one for every 15° of longitude. Any longitude line divisible by 15 (beginning at 0°) is a time-zone center, with the zone extending 7.5° to each side.
23. Four.
24. *Ante meridiem* (before noon); *post meridiem* (after noon).
25. No. State the 12 o'clock time as 12 noon or 12 midnight.
26. To have more hours of sunlight when most people are awake, which also conserves energy.

27. (a) Clockwise.
    (b) Counterclockwise.
28. When the shadow of the gnomon has a minimum value, it is pointing directly north, provided that the gnomon is north of the Sun's declination.
29. Daylight Saving Time begins at 2 A.M. on the first Sunday in April, and ends at 2 A.M. on the last Sunday in October.
30. The extremely brief interval 1.35×10^{-43} s following the Big Bang, when all forces were unified.
31. See the chapter Highlight for the answer.
32. The revolving of Earth around the Sun.
33. Date and latitude.
34. d
35. d
36. Autumnal equinox: when the Sun is directly over the equator and moving from north to south (about September 23). Vernal equinox: when the Sun is directly over the equator and moving from south to north (about March 31). Winter solstice: when the Sun is at its most southern point (December 23). Summer solstice: when the Sun is at its most northern point (about June 22).
37. The sidereal year is the time required for Earth to revolve around the Sun and reach the same position with respect to the stars. The tropical year is the time required for Earth to revolve around the Sun from one vernal equinox to the next.
38. Earth's tilt on its axis and its revolving around the Sun cause the Sun's radiation to be more direct on different parts of Earth's surface at different times of the year. This produces warm and cold regions at different latitudes at different times. Humans have divided these yearly changes into four divisions called seasons.
39. (a) About March 21 and September 22.
    (b) About June 21 and December 22.
40. They would be milder. A smaller tilt of Earth's axis would result in less change in heat received at the planet's surface.
41. The north star is on the northern horizon (0° altitude); 90°.
42. 39°. When the observer is in the Northern Hemisphere, the altitude of the north star is equal to the latitude of the observer.
43. c
44. d
45. Precession (of Earth) is the slow conical westward motion of Earth's axis of rotation caused by the gravitational forces of the Moon and Sun on Earth's equatorial bulge.
46. The very slow change in the position of stars with respect to the celestial poles.
47. 25,800 years.
48. d
49. a
50. The time for the Moon to go through its cycle of phases.
51. The most likely explanation for there being seven days in the week is that there are seven celestial bodies (Sun, Moon, and five visible planets) that move in relationship to the fixed stars.
52. Every four years.

53. Every four years, except century years not divisible evenly by 400; that is, there are 97 leap years in every 400 years.

54. Halloween originally marked the beginning of the winter season. Christmas originally celebrated the winter solstice and the beginning of the northward movement of the Sun.

ANSWERS TO CRITICAL THINKING QUESTIONS

1. The solar day would be shorter than the sidereal day.
2. Yes, at the equator. All meridians are one-half of a great circle. The only parallel that is a great circle is the equator.
3. A new day begins when 12 midnight arrives at the IDL. The new day moves westward as midnight moves westward. December 25 arrives first at and west of the IDL. When it is December 25 west of the IDL, it is December 24 east of the IDL. After celebrating Christmas on the west side of the IDL, cross the IDL traveling eastward. When 12 midnight of the 24–25 arrives at your new western longitude, you will have December 25 again.
4. The Sun appears to travel westward around the southern sky at varying altitude throughout the 24-h day of the summer solstice. The maximum altitude of the Sun will be 23.5° at 12 noon local solar time, and the minimum altitude will be about 23° at 12 midnight. The Sun never goes below the horizon on this date.

ANSWERS TO EXERCISES

1. 40°N – 20°S = <u>60°</u>

2. 105°E, 40°N
   ↑
   50°
   ↓
   • N.P.
   ↑
   50°
   ↓
   75°W, 40°N

   Answer: 50° + 50° = <u>100°</u>

3. $60° \times \frac{60 \text{ n.mi}}{1°} = \underline{3600 \text{ n.mi}}$

4. $100° \times \frac{60 \text{ n.mi}}{1°} = \underline{6000 \text{ n.mi}}$

5. West of the starting point.
6. East of the starting point.

7.

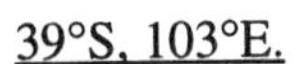

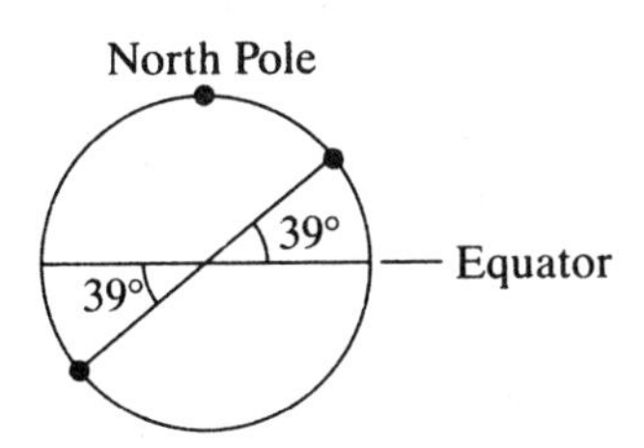

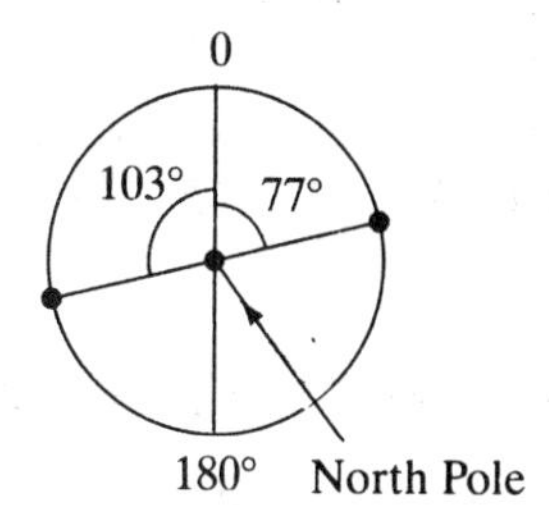

8. Solve in the same way as Problem 7. Answer is 36°S, 40°W .

9. 118°W – 80°W = 38 = 2 h, 32 min

   6:00 P.M. – 2 h, 32 min = 3:28 P.M. same date

10. 110°W – 70°W = 40° = 2 h, 40 min

    Midnight is occurring between the two meridians at 80°W.

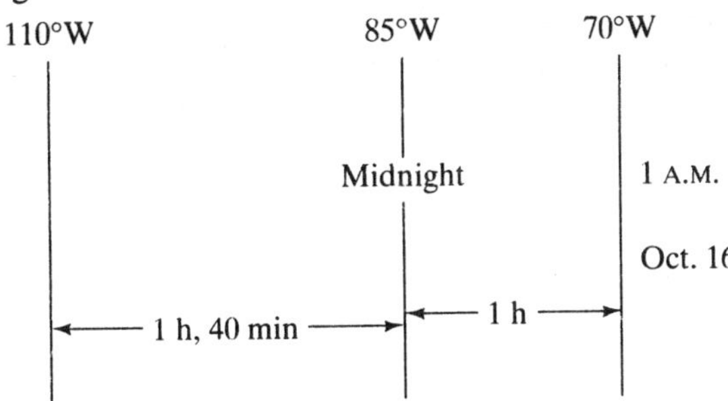

Answer: $t =$ 10:20 P.M. October 15

11. Work the same way as you did Problem 9. Tokyo is in the 135°E time zone. The answer is 3 A.M. the next day—November 27.

12. Work the same day as you did Problem 9. The same figure can be used again. The answer is 3 A.M. the next day—February 23.

13. The west coast is three time zones earlier than the east coast. 6 P.M. PST.

14. The west coast is three time zones earlier than the east coast. 4 P.M. PST.

15. Los Angeles is in the time zone centered on 120°W, and Moscow is in the time zone centered on 45° E. Using the figure at the right, start at 120°W and add an hour for every 15° of longitude as you go east. The answer is 9 P.M. on the same day—July 28.

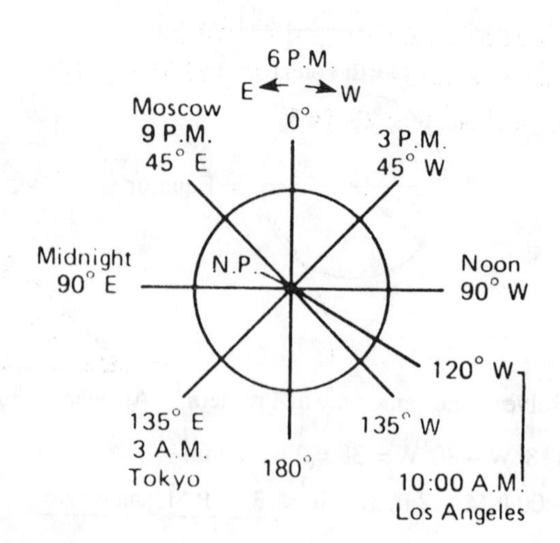

16. Atlanta is in the 75°W standard time zone. Milano is in the 15°E standard time zone. The two zones are 90° apart, or 6 h. The time in Milano is 7 P.M.

17. Use Table 17.1 to obtain the answer. Interpolate for daylight hours at 40°N between June 21 and December 22. There are 6 h over 6 months. July 21 will have approximately 14 h of daylight. Half of these hours are before noon. Seven hours before noon is 5 A.M.

18. Use Table 17.1 to obtain the answer. Interpolate for daylight hours at 40°N between June 21 and December 22. April will have approximately 13 h of daylight. Sunset will occur at 6:30 P.M.

19. (a) $Z = 42°$
so $A = 90° - 42° = 48°$.

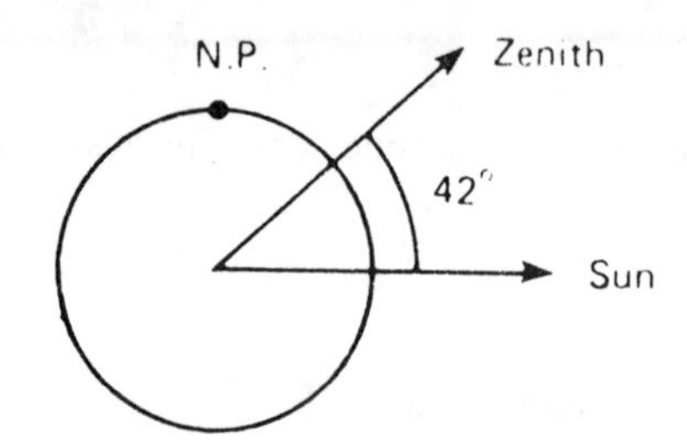

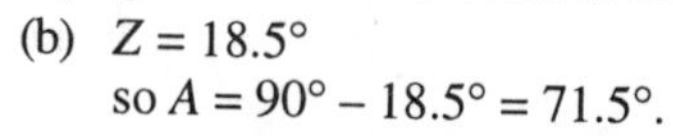
(b) $Z = 18.5°$
so $A = 90° - 18.5° = 71.5°$.

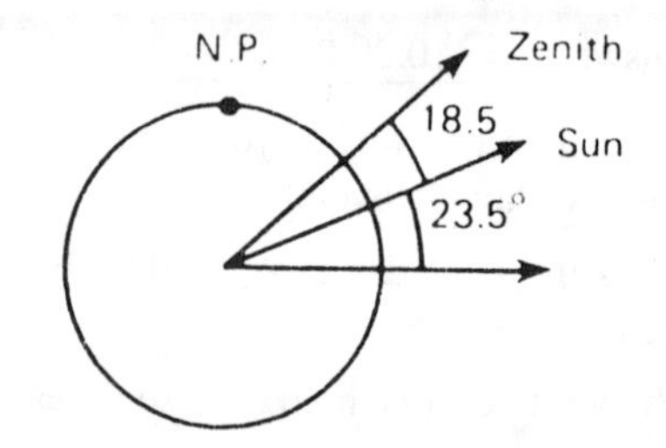

20. (a) $Z = 34°$
so $A = 56°$.

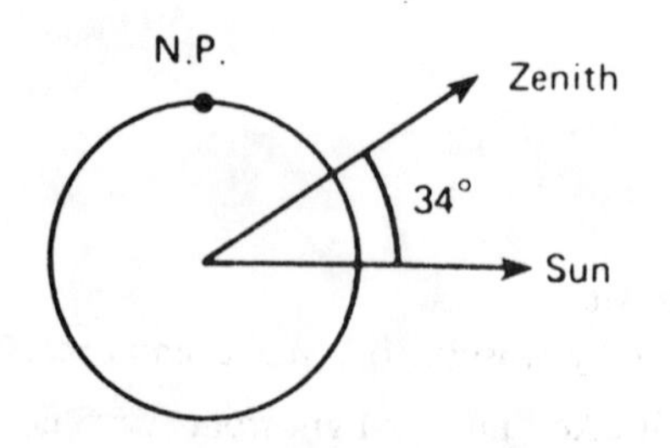

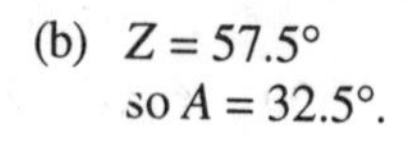
(b) $Z = 57.5°$
so $A = 32.5°$.

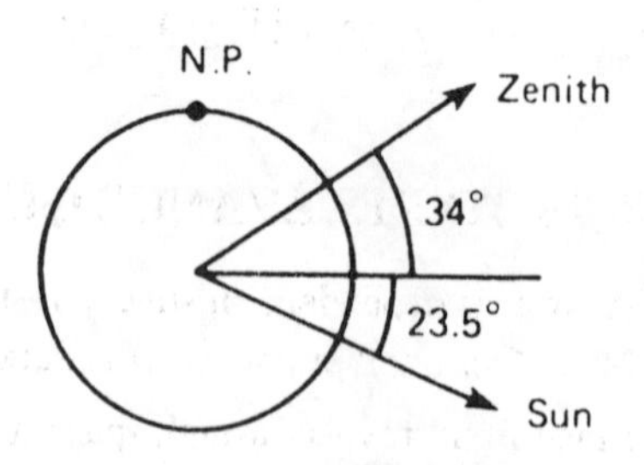

21. $A = 71.5°$, so $Z = 90° - 71.5° = 18.5°$.
Therefore, the latitude $= 18.5° + 23.5° = 42°N$.

22. A = 31.5°, so Z = 90° – 31.5° = 58.5°.
    Therefore, the latitude = 58.5° – 23.5° = 35°N.
23. On or about June 21, 74.5°.

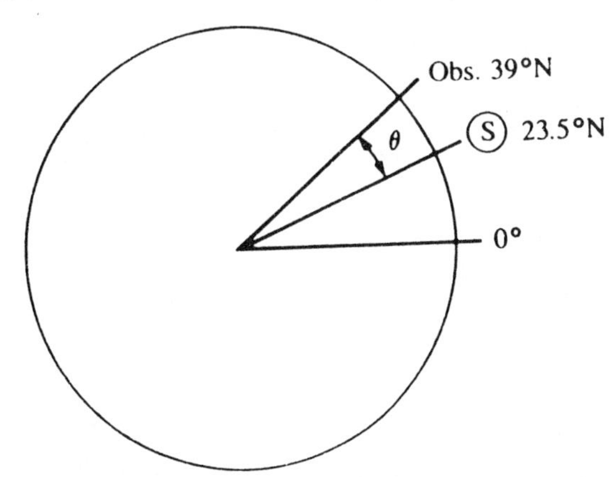

$$\theta = 39° - 23.5° = 15.5°$$
$$90° - 15.5° = 74.5° \text{ (maximum altitude)}$$

24. On or about December 22.

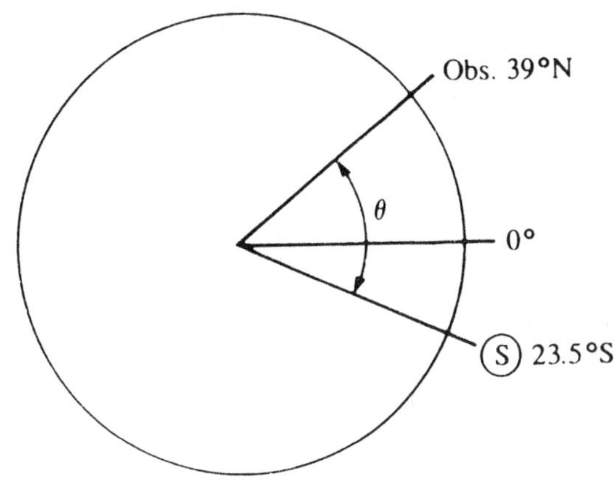

$$\theta = 39° - 23.5° = 62.5°$$
$$90° - 62.5° = 27.5° \text{ (minimum altitude)}$$

25. Yes. 74.5° – 27.5° = 47° and 23.5° × 2 = 47°.

ANSWERS TO RELEVANCE QUESTIONS

17.1 Answer is dependent on streets in student's hometown.
Note: Our number one street is Park Street, followed by Washington, Maple, and Lincoln.

17.3 Since light travels through space at 3×10^8 m/s, it takes time for light from stars and galaxies to reach the Hubble Space Telescope (HST). Therefore, the images received by HST are the way stars and galaxies appeared in the past. The HST can look back through 10 billion or more light-years of space.

17.4 (a) The typical heart rate is 72 beats per minute. At this rate, the heart beats 104,000 times each day. The process of regular breathing in and out occurs automatically approximately 12 to 22 times per minute.

(b) Sleep is regular and recurring about 7.5 h a night for a middle-aged adult. Eating is also regular and recurring, normally three times a day.

ANSWERS TO STUDY GUIDE QUIZ

Multiple-Choice Questions

1. c	2. d	3. d	4. d	5. d
6. a	7. a	8. a	9. a	10. d

Short-Answer Questions

1. Longitude is the angular distance east or west of the prime meridian that runs through Greenwich, England. Longitude is measured in degrees east (from 0 up to 180°E) or degrees west (from 0 to 180°W) of the prime meridian.
2. One nautical line is a unit of distance equal to the length of one arc minute on a great circle subscribed on Earth's surface. This means that one degree of arc equals 60 nautical miles, since there are 60 arc minutes per degree; 60 nautical miles is equivalent to 69 statute or land miles.
3. The second is currently defined as 9,192,631,770 oscillations of the 2*s* electrons of the cesium-133 atom as measured by a high-precision atomic clock.
4. Local solar time is the time for any location on Earth with respect to the current position of the Sun. When the Sun is directly on a particular meridian, the time is defined to be exactly 12 noon. For every 4 arc minutes that a person moves east or west of this location, the time changes by one minute, or for every 15 degrees, one hour. Since any travel across Earth's surface in an east-west direction would result in a continuous change in local solar time, a system of dividing the equatorial circumference into 24 equal 15° standard time zones has been adopted to avoid some of this confusion. Each standard time zone takes its time from the local solar time of the 15° prime meridian that runs essentially through the center of that time zone.
5. The International Date Line is the 180° E or W meridian, except for a small jog to go around a small land mass. If you cross the International Date Line from East to West, your calendar is advanced to the next day; if you cross from west to east, you lose a day.
6. Athens, Ohio, is in the Eastern Standard Time zone, so it takes its standard time from the position of the Sun relative to 75°, where the Sun is currently overhead, according to this question; that is, it is now noon local solar time at 75°W, and thus it is local noon for 7-1/2° to each side of this meridian. Since the actual longitude of Athens is 82°W, Athens is located 82 – 75 = 7° farther west, which is equivalent to 7 × 4 min = 28 min earlier. It is earlier because the Sun appears to travel from east to west and so has not yet been overhead in Athens on this day. This makes the local solar time in Athens 11:32 A.M. on the same date.
7. Daylight Saving Time is a way of providing sunlight longer into the evening hours, at the expense of having the Sun come up at a later time in the morning. In the summer months, some areas of the world set their clocks ahead of the normal standard time for their region by one hour. This means that the Sun is now highest in the sky at 11 A.M. and there will be one more hour of daylight left that day to work or play out-of-doors. It is believed that Daylight Saving Time can save energy and increases safety by reducing the occurrence of early-evening traffic accidents.

8. During the summer months, the Sun has a declination that places it north of the Equator and thus higher in the sky for longer periods of time than in the winter months, when it is lower in the sky and spends less time above the horizon. The Sun's rays also strike the Northern Hemisphere in a more concentrated fashion in the summer. All of this leads to warmer average temperatures in the Northern Hemisphere in the summer months.
9. The Sun's declination on December 21 is 23-1/2°S. This places the Sun overhead at 23-1/2°S and so leaves a zenith angle of 23-1/2° + 40°, or 63-1/2°, on this date.
10. The period of Earth's orbit around the Sun is not an even number of days in length, and so the extra quarter day builds up each year and pulls the dates of the spring and fall equinoxes out of synchronization with the accepted calendar. To get the calendar back into proper alignment with celestial events, one day must be added to the year every four years, which results in the occurrence of a leap year.

Chapter 18

The Moon

The *Clementine* spacecraft reached the Moon on February 19, 1994, and began its mission of mapping the Moon's surface rocks. The chapter-opening photograph shows the largest impact feature in the solar system, the South Pole–Aitkin Basin, which was known before the *Clementine* mission. The mapping data will provide scientists new knowledge about impact basins, other surface features, the lunar crust, the Moon's mineralogy, and the early history of the Moon.

This chapter deals with the origin of the Moon, the data gathered by the Apollo astronauts, the *Clementine* mission, the phases of the Moon, the tides, and eclipses of the Moon and the Sun.

Students should be given an outside assignment to observe the Moon for one month. Have them determine the time of meridian crossing, obtain the approximate maximum altitude using the astrolabe, and record the rising and setting times of the Moon.

The instructor should point out how the frame of reference is formed for the surface of the Moon. Many astronomy texts show the Moon reversed left to right, because telescope pictures are reversed. All photos in this text are oriented to show what someone on Earth would see if looking at the Moon with the unaided eye.

DEMONSTRATIONS

Use a model solar system with a Moon that shows the motion of Earth and the Moon and also a large Earth sphere and an 8-in.-diameter Moon that the instructor can move manually around the Earth sphere. Let the classroom wall be the Sun. Explain to the class the position of the Moon at new, first quarter, full, and last quarter phases. With the Moon at any of these positions, rotate Earth to show the rising and setting times of the Sun and the Moon. Demonstrate with a model sphere of Earth and the Moon, plus an incandescent lamp for the Sun, the positions necessary for solar and lunar eclipse.

ANSWERS TO REVIEW QUESTIONS

1. d
2. d
3. (a) 29.530588 days.
   (b) Same as revolution.
4. 3.476×10^3 km.
5. The surface gravity of the Moon is 0.165 that of Earth.

6. 2.4×10^5 mi, 3.8×10^5 km.
7. Rays are bright streaks on the Moon's surface that radiate from certain craters. Rills are crevasses or cliffs on the Moon's surface.
8. (a) These were mostly caused by meteorite bombardment.
   (b) Volcanic eruptions.
   (c) Pulverized rock thrown out when a crater was formed by a meteorite.
   (d) Moonquakes.
9. Crust, solid mantle, and the small iron-rich core.
10. The unique feature bout Earth and the Moon is that they are nearer in size than any other planet and its satellite, except for the minor planet Pluto and its satellite, Charon.
11. c
12. e
13. The oldest Moon rocks are 4.4 billion years.
14. The sister theory fails to explain why Earth has a great abundance of iron and water, whereas the Moon has very little iron and no water.
15. Volatile elements are those that are easily driven off by extreme heating. Refractory elements are those that are not easily vaporized.
16. (a) The Moon is more than 4.4 billion years old.
    (b) The age of Earth is thought to be some 4.6 billion years.
17. The Moon's average density is 3.3 g/cm^3. Earth's mantle has a density of about 3.5 g/cm^3.
18. The great impact theory proposes that a planet-size object, about the size of Mars, struck Earth a glancing blow 4.4 to 4.5 billion years ago. The impact ejected enough matter (most of it coming from Earth's mantle) into orbit to form the Moon.
19. The impact theory would account for the similar densities of Earth's mantle and the Moon and for the Moon's low abundance of iron. Also, the impact of a large object generates tremendous heat, which would drive off water and other volatile substances.
20. Rocks on the Moon have not been altered by chemical and physical weathering.
21. e
22. c
23. Counterclockwise as viewed from above the North Pole of Earth.
24. They are the same.
25. Sidereal period = 27.3 days (27 days, 7 h, 43 min, and 12 s).
    Synodic period (lunar month) = 29.5 days (29 days, 12 h, 44 min, and 3 s).
26. About 5° (5° 8' 43").
27. Earth's rotation produces moonrise and moonset.
28. (a) Every 29.5 days.
    (b) Every 29.5 days.
29. c
30. c

31. During a full moon, the Moon is on the opposite side of Earth from the Sun. Thus, when the Sun is high in the sky in the United States, the Moon is low, and vice versa.

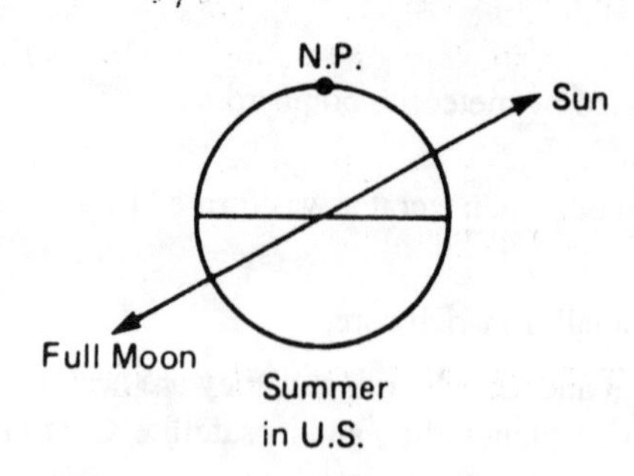

32. Every 29.5 days.

33. (a) 6 P.M.
    (b) 6 P.M.
    (c) 12 noon.

34. Because it is going around Earth every 29.5 days.

35. (a) First quarter.
    (b) Full moon.
    (c) Last quarter.

36. Waxing: more of the Moon's surface is seen each night.
    Waning: less of the Moon's surface is seen each night.

37. d

38. a

39. (a) The Moon is between Earth and the Sun.
    (b) Earth is between the Moon and the Sun.

40. Umbra is the complete dark shadow cast by an object. Penumbra is the semidark region.

41. (a) 18.6 years.
    (b) Westward.

42. Eclipses can occur only when the Sun, Moon, and Earth are in the same plane and in or near a direct line.

43. An eclipse of the Sun in which the Moon is too distant from Earth for the Moon to appear to cover the Sun completely; therefore, a ring of sunlight appears around the Moon at mid-eclipse.

44. Because the Moon is not usually in the ecliptic plane.

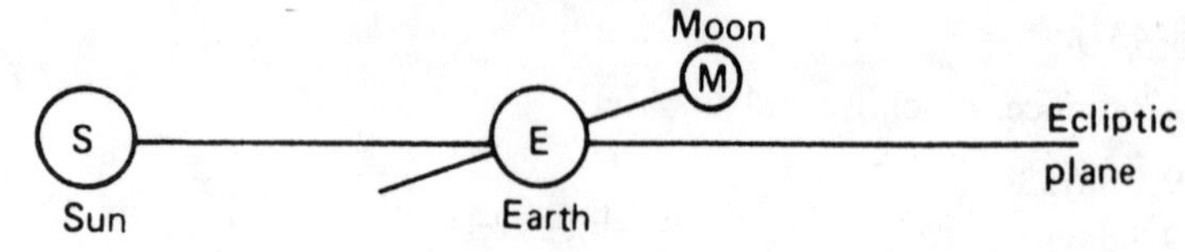

45. Because Earth's diameter is four times that of the Moon, it is easier for Earth to block out the Moon (lunar eclipse) than for the Moon to block out the Sun (solar eclipse).

46. b

47. d

48. The gravitational forces between Earth and the Moon and, secondarily, between Earth and the Sun.
49. The height of ocean tides is a function of the Moon's position in degrees latitude, the Sun's position with respect to the Moon and Earth, Earth's rotation, and the surface features of the boundary between land and ocean.
50. Spring: new and full. Neap: first quarter and last quarter.
51. See Section 18.6 for the answer to this question.

ANSWERS TO CRITICAL THINKING QUESTIONS

1. A crescent phase.
2. No. The relative positions of the Sun and Moon would remain the same.
3. West. The Moon revolves eastward around Earth; therefore, the Moon must enter Earth's shadow on the west side.
4. The Moon must rotate at twice its present period of rotation. The rotation period must be one greater than the revolving period.
5. It is the beginning of spring for the Northern Hemisphere. Spring for the Northern Hemisphere begins when the Sun appears to cross the equator going from south to north. The waxing crescent moon follows the new phase, which is on the meridian with the Sun.

ANSWERS TO EXERCISES

1. Approximately 100 N.
2. Approximately 20 lb.
3. Approximately 354 days (29.5×12).
4. Approximately 328 days (27.3×12).
5. (a) $35.5°N - 28.5°N = 7°$
   Maximum altitude = $90° - 7° = \underline{83°}$
   (b) <u>December 21</u>. The Moon is 180° east of the Sun.
6. (a) $36°N - 28.5°N = 7.5°$
   Maximum altitude = $90° - 7.5° = \underline{82.5°}$
   (b) <u>December 21</u>.
7. (a) $34°N - 28.5°N = 5.5°$
   Maximum altitude = $90° - 5.5° = \underline{84.5°}$
   (b) <u>March 21</u>. The Moon is 90° east of the Sun.
8. (a) $33.5°N - 28.5°N = 5°$
   Maximum altitude = $90° - 5° = \underline{85°}$
   (b) <u>September 22</u>. The Moon is 270° east of the Sun.

9.

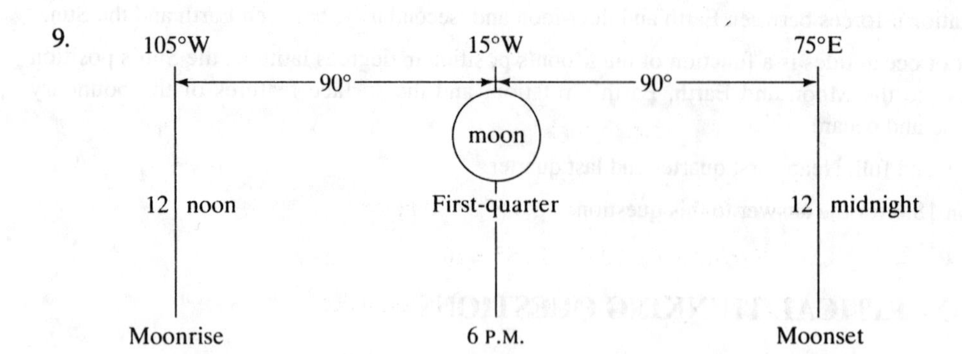

10.

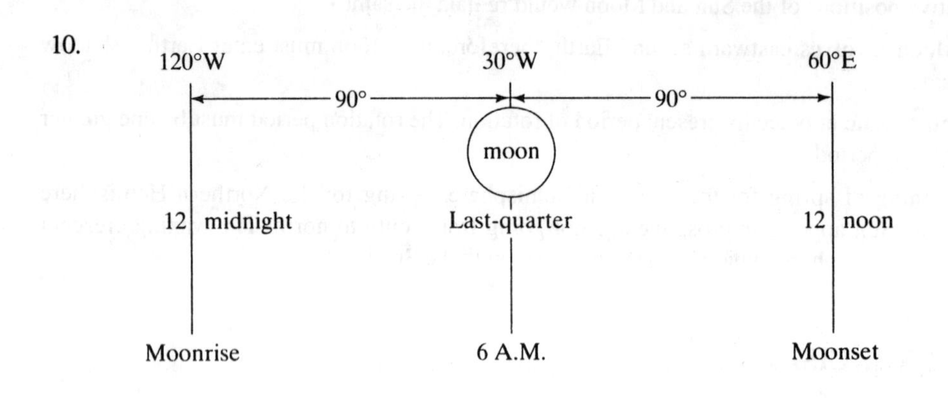

11. (a) The right side or west side is bright, because the Sun has just set in the west.
    (b) It would be the same phase in Australia and the west side would be bright, but this is the left side.

12. (a) The eastern side or the left side.
    (b) The same phase (last quarter), and the east or right side would be bright.

13. June 21 (summer solstice).

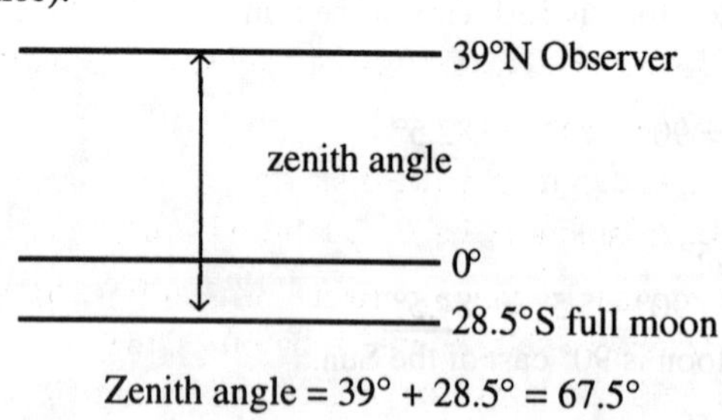

Zenith angle = 39° + 28.5° = 67.5°

90° – 67.5° = 22.5° (minimum altitude of full moon)

14. December 21(winter solstice).

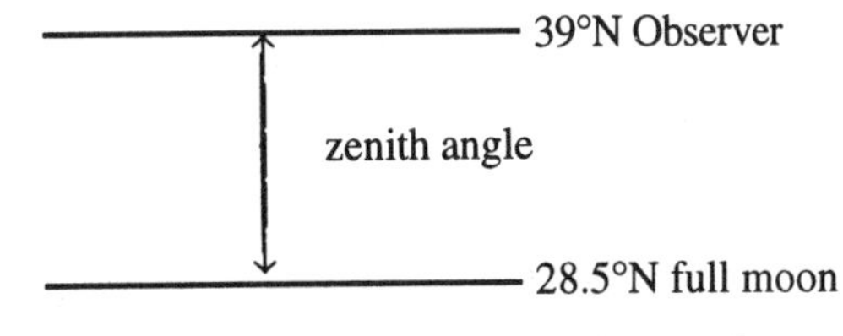

$$\text{Zenith angle} = 39° - 28.5° = 10.5°$$

$$90° - 10.5° = \underline{79.5°} \text{ (maximum altitude of full moon)}$$

15.

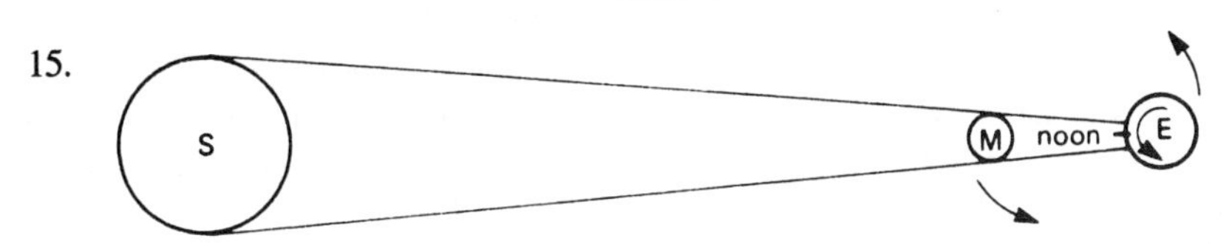

16.

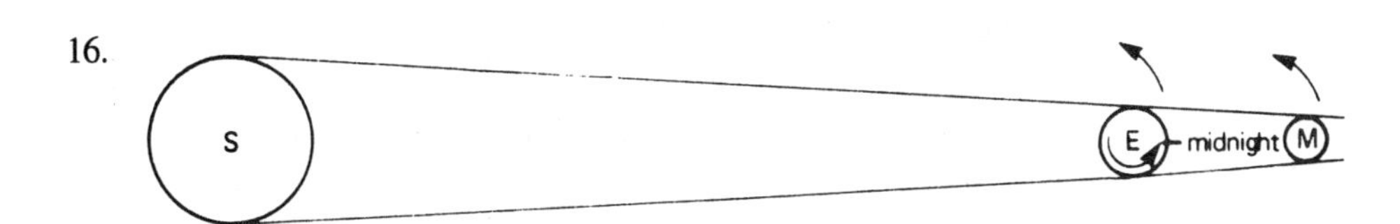

17. (a) High tides are 180° apart, so 99°E.
    (b) Longitudes 90° from 81°W and 99°E are 9°E and 171°W.

18. (a) 180° away, or 62°E.
    (b) 90° from 118°W and 62°E, or 28°W and 152°E.

ANSWERS TO RELEVANCE QUESTIONS

18.1 The largest features, which are plains and mountain ranges.

18.2 Government taxes, which we all pay, are used to support the space missions to the Moon.

18.3 The answer depends on the Moon's appearance and the student's outlook on life.

18.4 Religious holidays other than Easter.
Best fishing times (fishing tables are published yearly).
Agriculture planting times (planting tables are published yearly).

18.5 Smoked glass or photographic film will not protect the eyes from harmful ultraviolet radiation. An easy and simple way to observe a solar eclipse is by projecting the Sun's image through a pinhole in a piece of cardboard onto another piece of cardboard or any suitable screen. The distance at which a sharp image of a given size is formed depends on the size of the pinhole.

18.6 The height of the rise of the ocean tides would be less because the gravitational force between Moon and Earth would be less.

ANSWERS TO STUDY GUIDE QUIZ

Multiple-Choice Questions

1. b	2. d	3. a	4. d	5. c
6. a	7. b	8. d	9. a	10. a

Short-Answer Questions

1. Rays are streaks of light-colored material that extend outward from some of the Moon's craters. Rays are believed to be made up of pulverized rock that was thrown out of the craters when they were formed.
2. There is no atmosphere on the Moon, and so there is no rain or wind to cause erosion that would alter the lunar surface. This means that even ancient features have remained intact for billions of years. The erosion processes on Earth, however, tend to wipe out surface features such as impact craters in much shorter periods of time.
3. The preferred theory at present is the great impact theory, which proposes that the Moon was formed when a planet-sized object collided with Earth with sufficient force to throw enough matter from Earth's mantle into orbit to form the Moon.
4. The exact match between the orbital period of the Moon and its period of revolution means that the same side of the Moon's surface always faces Earth; that is, we always see the same side of the Moon facing us, and we never get to observe the back side of the Moon at all from Earth.
5. The phases of the Moon result from the fact that the half of the lunar surface that is lighted by the Sun does not always face Earth. Because of this, we see varying amounts of the lighted portion of the lunar surface over a month's time in such a way that the Moon begins as a small crescent sliver just after its new moon phase, waxes steadily through a fully illuminated phase at full moon, and then returns to new phase again in a 29.5-day cycle.
6. The eight lunar phases are new moon, waxing crescent phase, first quarter moon, waxing gibbous phase, full moon, waning gibbous phase, last quarter moon, and waning crescent phase. After the waning crescent phase, the cycle of lunar phases starts over with the new moon again.
7. The following chart shows the time of day each of the instantaneous phases of the Moon will be on an observer's overhead meridian.

New moon	12 noon
First quarter moon	6:00 P.M.
Full moon	12 midnight
Last quarter moon	6:00 A.M.

8. On December 21 the Sun will have a declination of 23.5°S, and so the full moon would be at 23.5°N if it were on the ecliptic at that time. It is possible for the Moon to be 5° farther north because of its ±5° variation from the ecliptic plane. This means that the maximum declination of the Moon could be 28.5°N. The zenith angle to the Moon would then be $38° - 28.5° = 9.5°$, which would make the maximum altitude $90° - 9.5° = 80.5°$ on December 21.
9. If the last quarter moon is on the ecliptic, it should be 90° ahead of the Sun, which rises at 6:00 A.M. on this day at all points on Earth. As 90° corresponds to 6 h of time difference, the last quarter moon should be rising 6 h earlier than the Sun, or at 12 midnight.
10. The Moon does not always lie directly on the ecliptic plane and can be as much as 5° above or below this plane when it is in new phase. This means that Earth, the Sun, and the Moon do not always line up so that the shadow of the Moon falls on Earth's surface to produce an eclipse. A solar eclipse can occur only when the new moon is on, or very near, ecliptic plane at this point in its monthly revolution around Earth.

Chapter 19

The Universe

The universe as viewed from planet Earth is composed of stars and star systems called galaxies. The stars and galaxies in turn form the myriad clusters that are observed throughout the depths of space. This chapter discusses these basic building blocks (stars and galaxies) and presents information on the structure, shape, form, age, and extent of the known universe.

The origin of the universe is slowly being understood, and the basic ideas behind our understanding are presented in this chapter. The birth, life, and death of stars are fairly well known, and basic information on one star (our Sun) is discussed.

The Hertsprung-Russell diagram should be explained and made familiar to the student. A take-home open-book quiz is an excellent way for the student to learn how the diagram is plotted and what it explains. This chapter also presents information on pulsars, black holes, and dark matter. The latest information on the detection of large structures of galaxies and how this information changes the view held by astronomers concerning the cosmological principle and the origin of the universe is also discussed.

DEMONSTRATIONS

Use a transparent globe to show star clusters and the motion of Earth and the Sun with respect to the stars. Color slides showing the spectrum of a nearby star and a star moving rapidly away from Earth are helpful in illustrating the Doppler shift. A variety of slides of celestial objects is available. A transparent model of the Milky Way galaxy is also available; it is useful in showing spatial relationships among the galaxy's spiral arms, the Sun, and major star clusters.

ANSWERS TO REVIEW QUESTIONS

1. e
2. b
3. A plasma is a gas of very rapidly moving positively charged nuclei and negatively charged electrons.
4. Hydrogen and helium.
5. (a) 15 million K.
   (b) 6000 K.
6. (a) A temporary cool region in the solar photosphere that appears dark.
   (b) Yes. See Section 19.1 for explanation.

7. The photosphere.
8. A radial flow of particles (mostly electrons and protons) from the Sun extending millions of miles into space.
9. The Sun's corona is visible only during a solar eclipse.
10. Nuclear fusion reactions inside its core.
11. Neutrinos have no mass (we think), have no charge, travel at the speed of light, and rarely interact. Photons have no mass, have no charge, travel at the speed of light, and readily interact with matter (e.g., protons and electrons).
12. b
13. c
14. a
15. The celestial sphere is the apparent sphere of the sky that has a very large radius centered on the observer.
16. Declination is the angular distance of a celestial object north or south of the celestial equator (+40°).
17. Right ascension is the angle measured eastward along the celestial equator from the vernal equinox to the hour circle passing through an object. An example would be 2 h 28 min.
18. The celestial prime meridian is the great circle passing from the North Celestial Pole to the South Celestial Pole perpendicular to the celestial equator and intersecting the celestial equator at the vernal equinox.
19. (a) The distance light travels in one year, which is 9.46×10^{12} km (5.87×10^{12} mi).
    (b) The distance away a star would be if it had a parallax of one second of arc.
    (c) The distance between Earth and the Sun, 1.5×10^{8} km (9.3×10^{7} mi).
20. The parsec is larger. One parsec equals 3.26 ly.
21. Seconds of arc. The distance to a star is one parsec when the stellar parallax is one second of arc.
22. Virgo and Leo. See Fig. 19.10.
23. Asterisms are familiar groups of stars that are part of a constellation or part of different constellations. The Big Dipper is an example.
24. d
25. b
26. A star is self-luminous sphere of gas.
27. Sirius has an apparent magnitude of –1.43 and is 8.7 ly from the Sun.
28. (a) Hydrogen and helium.
    (b) It is believed that a star's interior composition is about 60 to 80% hydrogen, 16 to 36% helium, and about 4% heavier elements.
29. A plot of absolute magnitude against temperature, spectral class, or color index for a group of stars.

30.

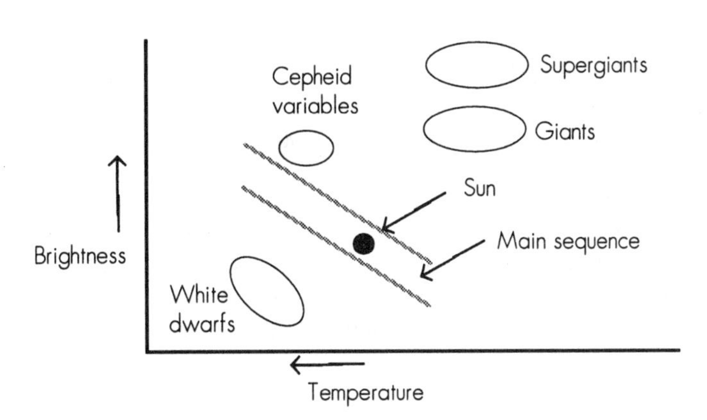

31. (a) Proxima Centauri
    (b) 4.3 ly
    (c) No. Alpha Centauri A, Alpha Centauri B, and Proxima Centauri are close companions and are located near 14 h right ascension and –61° declination.
32. A white star has the higher surface temperature.
33. Binary stars are double stars—i.e., two stars revolving about each other. No, but binary stars are very common in the universe.
34. Cepheid variables are pulsating variable stars of a class named after the prototype Delta Cephei. They are important to astronomers because they obey a period-luminosity relationship and are therefore useful in measuring distances.
35. Gravitational accretion, hydrogen burning, red giant, cepheid variable, white dwarf.
36. The star is moving away from the observer.
37. (a) A star of extremely high density composed almost entirely of neutrons.
    (b) A pulsating radio source believed to be associated with a rapidly rotating neutron star.
38. c
39. c
40. The collapse of a very massive body because of its attraction for itself.
41. A singularity is the term applied to the infinite-density point of a black hole.
42. The radial distance of the event horizon is equal to two times the universal gravitational constant times the mass of the body divided by the speed of light squared.
43. (a) A hypothetical object whose gravity is so strong that the escape velocity exceeds the speed of light.
    (b) Although there are some dark objects that are good candidates for black holes, and astronomers may believe that black holes exist, current experimental evidence offers no proof that they exist. The latest and best candidate is A0620-00 in the constellation Monoceros.
44. Search for binary systems in which one body is invisible and very massive, from which energetic X-rays are emitted.
45. The radius is directly proportional to the mass of the star and inversely proportional to c^2.
46. e
47. d
48. A galaxy is a very large group of stars held together by gravitational forces. Andromeda, Sombrero, and Milky Way.

49. (a) The way they appear in a photograph.
    (b) Elliptical, spiral, irregular.
50. The Milky way galaxy is of the spiral Sb type, contains about 100 billion stars, has a diameter of about 10^5 light-years, and has a thickness of 2000 ly in the region of the Sun.
51. (a) Local Group.
    (b) 21 known members.
    (c) M31 in Andromeda.
52. Isotropic means appearing the same in all directions. When we observe a large volume of space, we observe as many galaxies in one direction as in any other. Homogeneous means being uniform in properties throughout.
53. Dark matter is unobserved matter hypothesized by astronomers to explain why a cluster of galaxies exists as a gravitationally bound system.
54. A grouping of galaxies, the largest structure known at the present time, about 400 million ly long, 200 million ly high, and 15 million ly thick.
55. (a) See Section 19.5 for a complete answer.
    (b) Yes, recent observations of large groups of galaxies.
56. e
57. b
58. A starlike object considered to be a powerful energy source that exhibits a very large red shift; a shortened term for "quasistellar radio source."
59. (a) Blue.
    (b) –25.
60. (a) Quasars emit enormous amounts of electromagnetic radiation, and their spectral lines show an extremely large red shift.
    (b) Quasars appear small, blue, and starlike.
    (c) Quasars emit more radiation than stars, and they exhibit large red shifts.
61. e
62. e
63. See Section 19.7 for a complete answer.
64. (a) The cosmological red shift.
    (b) The 3-K cosmic microwave background radiation.
    (c) The mass ratio of hydrogen to helium of three to one in stars and interstellar matter.
65. See Section 19.7 for this answer.
66. The age of the universe can be calculated by taking the reciprocal of Hubble's constant.
67. (a) About 5 billion years.
    (b) About 15 to 20 billion years.
68. Because of the observed red shift of both near and distant galaxies.

ANSWERS TO CRITICAL THINKING QUESTIONS

1. The recessional velocity and the distance to a galaxy are needed to obtain a value for Hubble's constant. The recessional velocity is determined from the Doppler red shift in the galaxy's spectrum. Galaxies are members of rotating clusters. Therefore, in order to obtain the true recessional velocity of a relatively close galaxy, the observer must account for the rotational velocity when analyzing the Doppler red shifts, or blue shifts if the galaxy is rotating toward the observer. When a very remote galaxy is observed, the rotational velocity is small compared to the true recessional velocity, and the rotational velocity can be ignored. This gives a possible value for the recessional velocity. But the true distance to a very remote galaxy is difficult to obtain. Thus, at present, astronomers are unable to determine exactly both the recessional velocity and the distance.
2. The recessional velocity of a galaxy cannot exceed the velocity of light (3.00×10^8 m/s). Assume Hubble's constant to be 55 km/s/Mpc. Insert these values into Hubble's law ($v = Hd$) and solve for the distance.
3. If we assume that the Big Bang model of the universe is correct and that gravitational forces control the expansion of the universe, then the major factor that determines the future of the universe is density.
4. If we assume the Big Bang model of the universe to be correct, then it seems unreasonable to believe that our universe is one of a kind, because the Big Bang model implies a beginning and an end (Big Crunch). On the other hand, if there is no beginning or end for the universe, then it seems reasonable to believe that the universe is one of a kind.

ANSWERS TO EXERCISES

1. (a) Calculate 90% of Sirius' mass.
   $43 \times 10^{30}\text{ kg} \times 0.90 = \underline{3.87 \times 10^{30}\text{ kg}}$
   (b) Calculate 12% of the hydrogen.
   $3.87 \times 10^{30}\text{ kg} \times 0.12 = \underline{0.46 \times 10^{30}\text{ kg}}$
   (c) Convert one year to seconds. See inside back cover of textbook.
   $\underline{1\text{ y} = 3.16 \times 10^7\text{ s}}$
   (d) Calculate (N) the number of years required to convert 12% of the hydrogen into helium and energy.
   $2.0 \times 10^{12}\ \cancel{\text{kg}}/\cancel{\text{s}} \times 3.16 \times 10^7\ \cancel{\text{s}}/\text{y} \times N = 4.6 \times 10^{29}\ \cancel{\text{kg}}$
   $\underline{N = 7.3 \times 10^9\text{ y}}$
2. (a) Calculate 92% of the star's mass.
   $4.0 \times 10^{31}\text{ kg} \times 0.92 = 3.68 \times 10^{31}\text{ kg}$
   (b) Calculate 15% of the hydrogen.
   $3.68 \times 0.15 = 0.55 \times 10^{31} = 5.5 \times 10^{30}\text{ kg}$
   (c) Convert one year to seconds. See inside back cover of textbook.
   $1\text{ y} = 3.16 \times 10^7\text{ s}$
   (d) Calculate (N) the number of years required to convert 15% of the hydrogen into helium and energy. $3.0 \times 10^{13}\ \cancel{\text{kg}}/\cancel{\text{s}} \times 3.16 \times 10^7\ \cancel{\text{s}}/\text{y} \times N = 5.5 \times 10^{30}\ \cancel{\text{kg}}$
   $\underline{N = 5.8 \times 10^9\text{ y}}$
3. A parallax of 0.20 s implies a distance of 5.0 parsecs (see Problem 4); 5.0 parsecs × 3.26 light-years/parsec = <u>16 light-years</u>
4. Distance in parsecs = 1/0.20 = <u>5.0 parsecs</u>

5. $186{,}000\ \text{mi/s} \times 86{,}400\ \text{s/day} \times 365.25\ \text{days/y} = \underline{5.87 \times 10^{12}\ \text{mi}}$

6. $5.87 \times 10^{12}\ \text{mi} \times 1609\ \text{m/mi} = \underline{9.44 \times 10^{15}\ \text{m}}$

7. $680\ \text{kpc} \times \dfrac{3.26 \times 10^{3}\ \text{ly}}{1\ \text{kpc}} = \underline{2.2 \times 10^{6}\ \text{ly}}$

8. $4.3\ \text{ly} \times \dfrac{1\ \text{pc}}{3.26\ \text{ly}} = \underline{1.3\ \text{pc}}$

9. –3 is 10 magnitudes brighter than +7, and so it is $100 \times 100 =$ <u>10,000 times brighter</u> .

10. (a) 1 magnitude difference = <u>2,512 times brighter</u> .
    (b) 5 magnitude difference = <u>100 times brighter</u> .

11. $$R = \frac{2 \times 6.67 \times 10^{-11}\ \text{N-m}^2/\text{kg}^2 \times 15 \times 10^{30}\ \text{kg}}{[3.00 \times 10^{8}]^2\ \text{m}^2/\text{s}^2}$$

$$= 2.22 \times 10^{4}\ \text{m or } \underline{22\ \text{km}}$$

12. $$R = \frac{2GM}{c^2} = \frac{2 \times 6.67 \times 10^{-11}\ \text{N-m}^2/\text{kg}^2 \times 2.0 \times 10^{32}\ \text{kg}}{(3.00 \times 10^{8})^2 \text{m}^2/\text{s}^2}$$

$$= \underline{3.0 \times 10^{5}\ \text{m}}$$

13. (a) $100 \times 10^{9} \times 100 \times 10^{9} = \underline{10^{22}}$

14. $\dfrac{5 \times 10^{9}\ \text{y}}{2 \times 10^{8}\ \text{y}} = \underline{25}$

15. To obtain the answer to this problem in seconds, the distance in megaparsecs (Mpc) must be converted to kilometers (km) in order to cancel km out of the equation.

Solution:

$t = 1/H = 1.100\ \text{km/s/Mpc}$

$3.086 \times 10^{19}\ \text{km} = \underline{1 \times 10^{6}\ \text{pc}}$

$$t = \frac{1}{(100\ \text{km/s})/(3.086 \times 10^{19}\ \text{km})} = 3.086 \times 10^{17}\ \text{s}$$

$$\frac{3.086 \times 10^{17}\ \text{s}}{1} \times \frac{1\ \text{day}}{8.64 \times 10^{4}\ \text{s}} \times \frac{1\ \text{y}}{3.65 \times 10^{2}\ \text{days}} = \underline{9.78 \times 10^{9}\ \text{y}}$$

16. $$t = \frac{1}{\dfrac{75\ \text{km/s}}{3.086 \times 10^{19}\ \text{km}}} = \frac{3.086 \times 10^{19}\ \text{s}}{75} = 4.1 \times 10^{17}\ \text{s}$$

$$\frac{4.1 \times 10^{17}\ \text{s}}{1} \times \frac{1\ \text{day}}{86{,}400\ \text{s}} \times \frac{1\ \text{y}}{365\ \text{days}} = \underline{13.1 \times 10^{9}\ \text{y}}$$

ANSWERS TO RELEVANCE QUESTIONS

19.1 Energy is essential to maintain the interactions within and among Earth's four ecosystems—atmosphere (air), lithosphere (soil and rocks), hydrosphere (water), and biosphere (living organisms). A reduction of energy from the Sun by 0.5% would produce severe climate changes. Earth would enter an ice age and the ecosystem would change drastically, initiating conditions that the human body could not tolerate. Daily life would be unbearable, and millions of people would die.

19.2 The answer is dependent on student's awareness of the planets, Moon, and stars.

19.3 Yes. Some inhabitants on Earth's surface would see the stars briefly during an eclipse of one Sun.

19.4 It is impossible to know what takes place inside the event horizon of a black hole, because nothing can escape outward from the event horizon.

19.5 The Milky Way is the combined glow of millions of stars that appears as a band in the sky. We are located in the same plane as the stars. As Earth rotates and revolves, the orientation of the band changes. See the sky charts in the appendix.

19.6 Answers are dependent on the student's awareness of and interest in knowing the size, shape, and evolution of the universe in which he or she lives.

19.7 Since student can only engage in thought or reflection concerning the question, there should be a variety of answers.

ANSWERS TO STUDY GUIDE QUIZ

Multiple-Choice Questions

1. b	2. c	3. d	4. a	5. d
6. d	7. c	8. c	9. c	10. a

Short-Answer Questions

1. Any two of the following would be acceptable.

   Granules—hot spots, about 100 K hotter than the surrounding surface. Granules are a few hundred miles in diameter and last only a few minutes.

   Chromosphere—thin reddish crescent just above the photosphere that can best be seen for only a few seconds during a solar eclipse.

   Sunspots—patches of cooler material in the Sun's surface that can be up to several thousand kilometers in diameter and that appear dark against the hotter photosphere.

   Prominences—great eruptions at the edge of the Sun that are produced in conjunction with violent magnetic storms in the chromosphere. They appear as red streamers, loops, fountains, columns, or curtains against the dark background of space.

2. The Sun is powered by the fusion reaction in which hydrogen is converted into helium in the solar core, producing large amounts of heat and other forms of energy. This reaction is possible because of the high temperature and density within the Sun's core region.

3. The celestial sphere is an imaginary, transparent sphere on which all of the fixed stars are located and which rotates as a unit to produce the apparent movement of the stars around Earth. It is, of course, the rotation of Earth on its axis once every 24 h that is the real cause of this motion. The celestial equator on the celestial sphere is located directly above Earth's equator, and the North Celestial Pole is above Earth's North Pole. The Earth should be imagined as a tiny point at the center of the celestial sphere.

4. Constellations are groupings of stars as they appear from Earth. They tend to form recognizable patterns in the sky, and all stars are assigned to one of these pattern groups. The names assigned to these star groups can be traced back to the early Babylonian and Greek civilizations. Although the constellations have no physical significance in the actual structure of the universe, they are a convenient way of referring to certain areas of the sky as seen from Earth.

5. The red giant phase of a star's life comes after its main sequence lifetime, when most of the hydrogen in the star's core has been depleted and the star must switch to the use of helium as its core nuclear fuel.
6. Star spectral types are presented by the letters O, B, A, F, G, K, and M, which relate not only to the spectral appearance of the emitted light from the star, but also to the star's surface temperature, with O being the hottest and M the coolest. Our Sun has been determined to be a G-class star.
7. A supernova is an extremely violent explosion that occurs when a massive star has run out of core nuclear fuel and its gravitational pull collapses it into a neutron star. As the rapidly collapsing matter in the star suddenly encounters the newly formed neutron shell, a gigantic rebound of energy occurs that throws off much of the star's mass in a tremendous explosion.
8. An H-R diagram is a plot of the stars in a specific star cluster as a function of their absolute magnitude and their surface temperature. On this plot, the main sequence stars form a diagonal pattern from upper left, where the heaviest, brightest stars are plotted, to the lower right, where cooler, lighter stars are found. Red giant stars are always found in the upper right-hand corner of the diagram, above the main sequence stars, and white dwarf stars are below and to the left of the main sequence region. (See Figure 19.11 in the textbook.)
9. Galaxies are classified by their appearance as seen from Earth. The three main classes into which galaxies are fitted are elliptical, spiral, and irregular.
10. The cosmological red shift is a changing of the spectral line frequencies of the light from distant galaxies toward the low-frequency (red) end of the spectrum. This shift is an indication that the universe is expanding, and that all parts of the universe are continuously getting farther apart.

Chapter 20

Minerals and Rocks

The fundamental principle underlying most geology is simply that the processes occurring at present on Earth have occurred throughout geologic time. Thus ancient rocks can be interpreted with respect to present processes.

The composition of rocks is expressed in terms of minerals, which are the building blocks of rocks. Our study of geology begins with the study of minerals.

Mineral classification is based on physical and chemical properties. The physical properties distinguish different forms of minerals with the same composition. In addition, physical properties provide a convenient means of mineral identification.

Although the number of common minerals is small—fewer than 20 make up over 95 percent of the rocks in Earth's crust—the student may find the number (and names) overwhelming. As a result, only a few minerals are examined in detail. Among these is silicon dioxide (SiO_2) which occurs abundantly as quartz, sand, and many other minerals.

Following the basic information concerning minerals and igneous rocks, we consider the igneous activity that produces the rocks that form Earth's crust.

DEMONSTRATIONS

The best demonstrations for this chapter are representative mineral and rock collections that students can examine and analyze. Also, use rock charts that illustrate types, characteristics, identifying features, and the interrelationships of various types of rock. A chart showing the rock cycle should be a continuous exhibit in the classroom and laboratory during the teaching of geology. The chart on the chemistry of rocks is a good demonstration exhibit.

ANSWERS TO REVIEW QUESTIONS

1. d
2. b
3. d
4. Silicon is an element. It has an atomic number of 14 and a relative mass of 28, and is classified as a nonmetal. Silicates are minerals that have the oxygen-and-silicon tetrahedron as their basic structure.
5. A mineral is a naturally occurring, crystalline, inorganic substance (element or compound) that possesses a fairly definite chemical composition and a distinctive set of physical properties.

6. Percentage by mass: oxygen 46.5%, silicon 27.5%.
7. The silicon-oxygen tetrahedron is a covalently bonded structure composed of one silicon atom surrounded by four oxygen atoms. The structure is the basic building block of silicate minerals. See Fig. 20.2a for a sketch of its structure.
8. (a) Calcite ($CaCO_3$).
   (b) Galena (PbS).
   (c) Gypsum ($CaSO_4 \cdot H_2O$).
   (d) Halite (NaCl)
9. Lead (galena), iron (hematite), zinc (sphalerite).
10. Hardness, cleavage, color, streak, luster, crystal habit, fracture, tenacity, magnetism, fluorescence, and phosphorescence.
11. e
12. a
13. A rock is defined as a solid, cohesive, natural aggregate of one or more minerals.
14. Igneous rock, which solidifies from hot molten material called magma.
    Sedimentary rock, which can form in three ways:
    (1) The lithification of any preexisting sediment.
    (2) The precipitation of a mineral from a solution.
    (3) The consolidation of plant or animal remains.
    Metamorphic rock, which has been transformed from preexisting rock by high temperature or pressure or both.
15. See Fig. 20.7 for the answer to this question.
16. Magma is hot molten liquid located deep within Earth and composed of rock-forming materials.
17. Basaltic magma is a term applied to the hot melt that solidifies into a huge mass of rock called basalt.
18. See Fig. 20.6 for the answer to this question.
19. d
20. d
21. An igneous rock is one formed by the crystallization of molten magma.
22. The molten material is known as magma as long as it is beneath Earth's surface but becomes lava if it flows on the surface.
23. Intrusive rock is igneous rock that formed below Earth's surface. Extrusive rock is igneous rock formed outside Earth's crust. See Section 20.3 for properties.
24. Radioactivity, which is the spontaneous decay of certain unstable atomic nuclei.
25. Composition and texture.
26. Granite is high in silica; basalt is low in silica.
27. d
28. e
29. e
30. An igneous rock is discordant if it cuts across the grain of the surrounding rock and concordant if it is parallel to the grain.
31. Batholith.

32. Dikes are discordant rocks formed from magma that has filled fractures that are vertical or nearly so. A sill has the same shape as a dike but is concordant rather than discordant.
33. Chemical composition and temperature.
34. The outer rim of the Pacific Ocean is marked by a ring of volcanoes known as the Ring of Fire.
35. A "hot spot" is a fixed source of heat beneath one of Earth's lithospheric plates.
36. See Section 20.4 for the answer to this question.
37. Caldera is a term applied to a roughly circular, steep-walled depression in a volcano that may be up to several miles in diameter.
38. e
39. d
40. b
41. Common characteristics of sedimentary rocks are color, rounding, sorting, bedding, fossil content, ripple marks, mud cracks, and footprints.
42. Conglomerate, from gravel; shale, from mud; limestone, from shell fragments; sandstone, from sand.
43. Bedding—or stratification, as it is also called—is the layering that develops at the time the sediment is deposited.
44. Lithification is the transformation (compaction or cementation) of sediment into a clastic sedimentary rock.
45. Clastic rock is a sedimentary rock composed of fragments of preexisting rock. Organic rock is a sedimentary rock composed principally of the remains of plants and animals. Chemical rock is a sedimentary rock composed of chemical sediments that were precipitated from water by either an inorganic or an organic medium.
46. a
47. d
48. d
49. d
50. Metamorphic rocks are those that have been changed under the influence of very high temperature and/or pressure deep beneath Earth's surface.
51. See Section 20.6 for the answer to this question.
52. Foliation is the ability of a metamorphic rock to split along a smooth plane.
53. See Table 20.6 for the answer to this question.

ANSWERS TO CRITICAL THINKING QUESTIONS

1. Igneous rocks contain different minerals that have different melting temperatures.
2. Magma originates below Earth's crust at depths as great as 100 mi (161 km).
3. Physical disintegration and chemical decomposition of ancient rocks have taken place.
4. The answer depends on the location of the student.

ANSWERS TO RELEVANCE QUESTIONS

20.1. Pencil: wood and graphite
Finger ring: metal with gem
Wall plasterboard: paper and gypsum

20.2 Building materials and graveyards.

20.3 Granite, whose decomposition makes it more durable than limestone or marble.

20.4 As of this writing (4–3–96), Soufrire Hills, Montserrat, West Indies.

20.5 Knowledge from laboratory experiments can be used to identify layers of coal, conglomerate, limestone, sandstone, shale, and others.

20.6 Dark slate for school blackboards.
Flooring and roofing for buildings.
Patio slate for landscaping.
The best billiard tables for recreation.

ANSWERS TO STUDY GUIDE QUIZ

Multiple-Choice Questions

1. b	2. d	3. c	4. a	5. a
6. c	7. d	8. d	9. c	10. b

Short-Answer Questions

1. Feldspar has a chemical composition of sodium, calcium, or potassium in their aluminum silicate forms. It is gray-white to pink-white, and has a vitreous luster, a definite cleavage, a hardness of 6, and a specific gravity between 2.5 and 2.8. (See Table 20.2 in the textbook.)
2. Tenacity is the ability of a mineral to hold together. Some minerals are tough and durable; others are fragile and brittle. The tough and durable ones are said to possess tenacity.
3. Sedimentary rock forms at the surface of a planet or moon and can originate in three ways:
   (1) The conversion of preexisting sediments into rock.
   (2) The precipitation of minerals from solution into rock.
   (3) The consolidation of plant and/or animal remains into rock.
4. The grain size (texture) of igneous rock depends primarily on the rate at which molten rock cools. Slowly cooling magma gives the mineral grains in the sample time to grow, and, as a result, large crystals or grains form in the rock. The primary factors that determine the cooling rate of molten rock are the location, volume, and shape of the cooling rock sample. Large, massive igneous rock samples that have formed deep beneath the surface will have the largest grain size.
5. Both dikes and sills form when molten rock fills fractures in existing rock layers and then cools into igneous rock. The general shape of either is dependent on the fracture that it fills, so that such formations usually have a thin, platelike or tabular shape. The difference between them is that dikes are discordant; that is, they form across and perpendicular to existing rock layers, and this means that they generally lie in a basically vertical orientation. Sills, on the other hand, form in horizontal fractures that are generally concordant; that is, they form parallel to the existing rock layers around them.

6. Viscous lava tends to clog the throats of volcanoes; this can lead to great pressure build-ups, resulting in violent explosions when the plugs are finally blown free or the surrounding rock fractures. Nonviscous lava flows can pour large quantities of molten rock onto the surface without violent eruptive activity because the flow is generally smooth and continuous.
7. The Ring of Fire is an area surrounding the Pacific Ocean that is characterized by strong volcanic activity. This region of volcanic activity shows a marked correlation with plate boundaries, where subduction of surface crust can produce volatile magma formation below Earth's surface.
8. A caldera is a roughly circular, steep-walled depression formed as the result of the collapse of a volcanic chamber. This generally happens after the level of magma in a volcano recedes, leaving a partially empty chamber that collapses as a result of overlying debris from previous eruptions.
9. Thermal metamorphism is the change in the existing rock structure brought about by the proximity of heat, but under low-pressure conditions. This can occur when a molten body of hot magma works its way toward the surface, subjecting the surrounding rock to high temperatures that cause the metamorphism of these adjacent rock layers.
10. Slate is the metamorphosed form of shale or tuff. It is composed primarily of mica and quartz and shows excellent foliation and very fine grain. (See Table 20.7 in the textbook.)

Chapter 21

Structural Geology

Seismology is the study of vibrations and the resulting waves in Earth caused by earthquakes or human-caused explosions. Earthquakes are the result of sudden disruptions in Earth, and the resulting waves provide some clues about the composition and structure of our planet.

This chapter discusses information concerning Earth's internal composition and structure and the methods geologists use to obtain the information. This is followed by a discussion of the theories of continental drift, seafloor spreading, and plate tectonics.

The concepts of the theory of place tectonics are surprisingly simple, but the consequences are far-reaching. The theory provides a self-consistent view of the interrelatedness of geologic processes; and, in an effort to give the student a contemporary view of geology, the concepts of plate tectonics presented in this chapter will be used to explain various geologic phenomena in Chapter 22.

For this reason, the instructor should give students a well-grounded explanation of the theory of plate tectonics. A brief historical development of the theory comes from the study of continental drift and seafloor spreading, which provides justification for the idea of "moving" plates. The relative motions of plates should be emphasized for later references to the important geologic processes that occur along plate boundaries.

However, it should also be emphasized that the theory of plate tectonics is only a theory. Although it is grounded in recent geologic experimental evidence, plate tectonics is a relatively new theory and must stand many future tests of the scientific method.

DEMONSTRATIONS

Sources of teaching aids such as demonstration equipment and supplies, audio-visual material, and publications of use to instructors are listed in the Teaching Aids section at the end of this *Guide*. Films, videotapes, slides, and transparencies are helpful in teaching this chapter.

ANSWERS TO REVIEW QUESTIONS

1. e
2. d
3. d
4. e
5. Earthquakes may be caused by explosive volcanic eruptions or even explosions caused by humans, but most appear to be associated with movements in Earth's crust.

6. The focus of an earthquake is the point or region of the initial energy release or slippage. The epicenter is the location on Earth's surface directly above the focus.
7. Surface waves and body waves. Body waves are subdivided into P waves and S waves.
8. P waves are longitudinal compressional waves. S waves are transverse waves. S waves can travel only through solids. P waves can travel through any kind of material. See Figure 21.4.
9. (a) The Richter scale gives an absolute measure of the energy released by calculating the energy of seismic waves at a standard distance.
   (b) The Mercalli scale describes the severity of an earthquake by its observed effects.
10. Tsunamis are huge ocean waves coming ashore; they are commonly and incorrectly referred to as tidal waves, although they have no relation to tides.
11. d
12. d
13. d
14. Wegener's theory of continental drift is supported by (1) the similarities in biological species and fossils found on various continents; (2) continuity of geologic structures such as mountain ranges, and the distribution of rock types and ages; and (3) glaciation in the Southern Hemisphere.
15. The present mechanism for continental drift is seafloor spreading due to convection currents of subterranean molten materials. Wegener's explanation for continental drift was unacceptable because Earth's rotation could not supply the force necessary to overcome the measured strengths of rocks.
16. Remanent magnetism refers to the magnetism of rocks resulting from a special group of minerals called ferrites.
17. The formation of magnetic anomalies at the mid-ocean ridges. Lavas that form the ocean floor cool at the ridges, and they are magnetized by Earth's magnetic field as they solidify. Close examination of the magnetic intensity on each side of the mid-ocean ridge shows a mirror-image relationship.
18. A few centimeters each year.
19. c
20. b
21. The lithosphere is the solid outer shell of Earth; it includes the crust and some of the upper mantle. The asthenosphere is the region below the lithosphere that is hot enough to be easily deformed and is capable of internal flow.
22. A huge slab of lithosphere material.
23. Divergent, convergent, and parallel motions. See Figure 22.17.
24. The region where one plate is deflected downward and beneath another plate.
25. Volcanic activity comes from the rising of molten materials from subduction zones, and earthquakes result from energy released as the solid plates move in contact with and relative to each other.
26. African, North American, Pacific, Eurasian, Indian (or Australian), and Antarctic plates.

ANSWERS TO CRITICAL THINKING QUESTIONS

1. Seismic vibrations, tsunamis causing coastal flooding, fire causing extensive damage, rockslides.
   Do not build structures in fault zone regions.
   Design structures that are earthquake-resistant.
   Better earthquake prediction.
2. Convection currents in the asthenosphere; gravity pulling cold dense lithosphere down into the asthenosphere.
3. $$\frac{\frac{5000\ \cancel{\text{km}}}{1}\times\frac{1000\ \cancel{\text{m}}}{1\ \cancel{\text{km}}}\times\frac{100\ \cancel{\text{cm}}}{1\ \cancel{\text{m}}}}{2.0\ \frac{\cancel{\text{cm}}}{\text{y}}} = 2.5\times10^{8}\ \text{y}$$

   Each plate moves 2.0 cm/y in the opposite direction; therefore, the answer is

   $\frac{2.5}{2}\times10^{8}$ y = about 125 million years.
4. The absolute motion of seafloor spreading can be determined by dating rocks at different distances from the spreading ridge and dividing the distance traveled by the rock's age, which is the time it has taken the rock to travel that distance from the ridge where it was formed.

ANSWERS TO RELEVANCE QUESTIONS

21.1. Some students will give a positive answer to this question. Others will fail to respond.

21.2. Answers will vary from the fit of coastal areas of the South American and African continents, to fossil evidence, to ancient climatic similarities, to rock types and structures.

21.3. In North America, the North American Plate, and the projected movement is westward.

ANSWERS TO STUDY GUIDE QUIZ

Multiple-Choice Questions

1. b 2. d 3. c 4. a 5. d
6. d 7. d 8. a 9. a 10. c

Short-Answer Questions

1. The energy necessary for the violent movement of adjacent surface plates is stored in the elastic deformation of the rock along these boundaries. When this movement finally occurs, the released energy produces surface and body waves that are characterized by catastrophic seismic activity known as an earthquake.
2. The modified Mercalli earthquake scale measures the physical effects of an earthquake on people and their property. It is based on actual observations of the damage done by the earthquake.
3. A tsunami is a huge wave that is generated by the energy released during an earthquake and can travel across the ocean at speeds up to several hundred kilometers per hour. The disastrous effects of a tsunami come when this gigantic wave, sometimes over 15 ft in height, crashes into shore, producing immense destruction of property and sometimes even loss of human life.
4. The composition of Earth divides it into three layers; the crust, the mantle, and the core. The innermost of these, the core, has an average density of about 10 g/cm^3, which suggests a metallic composition (probably iron and nickel). The central 1200 km (radius) of Earth's core is probably

solid. The inner core is surrounded by a molten but highly viscous "liquid" outer core about 2250 km thick.

5. The lithosphere is the outermost crust of Earth; it extends downward to a depth of about 80 km. All of Earth's crust and the uppermost portion of the mantle are included in this designation. The lithosphere is rigid, brittle, and relatively resistant to deformation.

   The asthenosphere extends from the bottom of the lithosphere to a depth of about 700 km. This layer is more plastic and plays an essential role in the processes of isostatic adjustments, continental drift, and seafloor spreading.

6. The experimental verification of seafloor spreading came from deep-sea explorations using ship-mounted drilling rigs. Drilling samples showed that the layers of sediment on the seafloor were thinner near the center of mid-oceanic ridges. This indicates that these layers were recently deposited in newly formed crust as the ocean floor was pushed apart by upwelling magma from inside Earth's mantle. In addition, studies of the remanent magnetism along these ocean ridges show magnetic bands within the crust that have moved away from the ridges, again supporting the theory of seafloor spreading.

7. The plates on either side of a transform boundary slide past each other in a very slow but persistent manner in such a way that there is little or no destruction of the lithosphere itself. Sudden releases of elastic energy along these boundaries, when the plates finally slip, can result in strong earthquake activity in these regions.

8. Convection cells result from the heating of magma deep beneath Earth's mantle. The magma expands and subsequently rises toward the surface because of its reduced density. This upwelling magma tends to spread out under the surface plates, dragging them along with the flow and producing crustal plate movement in a process known as plate tectonics.

9. When lithospheric plate material is dragged down into a subduction zone, it becomes heated and quite often melts, producing molten magma and volatile gases that work their way back to the surface. This process results in volcanic activity along the subduction zone that can cause earthquakes and often leads to the formation of volcanic mountain ranges.

10. The Appalachian Mountains were formed as the result of the convergence of two large lithospheric plates that produced subduction with related volcanism and folding of the crust, and a later collision that thrust one section of crust up and over the North American coastal sediments. Continued collision processes caused additional compression, which resulted in more deformation and metamorphism of the rock throughout the Appalachian region.

Chapter 22

Isostasy and Diastrophism

The hypothesis of isostasy—the concept of a floating lithosphere in gravitational balance—is fundamental to understanding changes in the major topographical features of Earth (continents and ocean basins).

Diastrophism is a general term used by geologists for all types and modes of deformation of Earth's crust. Some deformations, especially most folding and faulting, take place slowly. Others, such as volcanic eruptions and earthquakes, bring about sudden changes.

Some of the great mountain chains—the Alps, Andes, Appalachians, and Himalayas, with their fascinating folded rocks—confirm the processes of faulting and folding of Earth's crust. Mountains are also formed by such processes as volcanic eruptions, where there is an outpouring of lava and tephra, or the uplifting of huge lithospheric plates when plate boundaries collide.

This chapter discusses the concept of isostasy and the processes involved in mountain building.

DEMONSTRATIONS

Geology models and demonstration kits showing geological features in three dimensions are very useful in teaching this chapter. The models and kits are available from scientific equipment suppliers. See the Teaching Aids section at the end of this *Guide*.

ANSWERS TO REVIEW QUESTIONS

1. d
2. b
3. The concept of isostasy states that a tendency toward equilibrium, operating in the components of Earth's crust, works to bring all rock masses into balance.
4. The continental crust differs fundamentally from the ocean crust in density.
5. Continents and ocean basins stand at different elevations because they are composed of rocks of different density. The continents are less dense.
6. The study of seismic waves from earthquakes, which indicate the presence of a partially molten material immediately below the lithosphere.
7. Under the great weight of ice, the rock surface subsides until isostasy is reestablished.
8. When great weight is removed from a rock surface by erosion, the rock mass will rise because the downward forces on the rock have decreased.

9. b
10. d
11. An anticline is a rock fold with downward sloping on both sides of a common crest. A syncline is a rock fold with upward sloping on both sides of a common trough.
12. Normal faulting: divergent plate boundaries.
    Reverse faulting: convergent plate boundaries.
    Strike-slip faulting: parallel moving plates.
13. Strike-slip faulting.
14. A normal fault occurs as the result of expansive forces that cause the overlying side of the fault to move downward relative to the side beneath it. A reverse fault occurs as the result of compressional stress forces that cause the overlying side of the fault to move upward relative to the side beneath it.
15. See Section 22.2 for the answer to this question.
16. The fault plane is the surface along which the rocks move relative to each other when a fracture occurs.
17. b
18. c
19. Diastrophism refers to the series of processes by which the major features of Earth's crust are formed and changed because of relative changes in the position and deformation of rocks.
20. Volcanic mountains have high peaks, cone shapes, and cuplike craters at their summits. The Cascade Mountains in Washington and Oregon are an example. See Fig. 22.8. Fault-block mountains are built when large rock units are displaced along high-angle normal faults. Mount Whitney in California is an example. See Fig. 22.9. Folded mountains are characterized by folded rock starts. The Appalachian Mountains are an example. See Fig. 22.10 for another example.
21. Volcanic mountains are located above the subduction zones of plate boundaries. See Fig. 21.20. Folded mountains are formed when a descending oceanic plate pushes the depositional basin strata against the continental plate or when two continental plates collide. Fault-block mountains are generated when normal faulting takes place at high angles. The upfaulted block rises sharply above the downfaulted block.
22. Depositional basins are long, narrow, subsiding ocean troughs containing large accumulations of sediment.
23. (a) Collision of the Indian and Eurasian Plates.
    (b) Collision of the African and Eurasian Plates.
    (c) Collision of pre-Pangaea plates that formed Pangaea.
24. The convergent boundary.

ANSWERS TO CRITICAL THINKING QUESTIONS

1. The two most important features of Earth's surface are oceans and continents.
2. Ocean basins are composed mainly of basalt. Continents are composed mainly of igneous and metamorphic rocks of granite composition. Granite continents stand higher because they are less dense and thicker than the ocean basins.
3. Faulting and folding depend on the strength of the rock, which is a function of the physical properties of the rock plus the pressure and temperature environment. Folding takes place more easily at depth, whereas faulting is prevalent near the surface.

4. Continents persist because, as they erode away, they become lighter and are buoyed upward.

ANSWERS TO RELEVANCE QUESTIONS

22.1. Most students will probably state a direct relationship, since present changes in geological processes give no indication of the age of Earth.

22.2 Anticlines and synclines are easily identified. Rock identification depends on students' ability to remember laboratory identification of rock samples.

22.3 The block of ice will sink when additional weights are added and rise when weights are removed. The elevation of a mountain reacts in a similar way.

ANSWERS TO STUDY GUIDE QUIZ

Multiple-Choice Questions

1. d	2. c	3. d	4. b	5. b
6. a	7. d	8. a	9. b	10. c

Short-Answer Questions

1. Since modern geologists cannot go back into history to study the processes that formed Earth's geologic features, it is necessary to base the science of geology on the premise that the present is the key to the past. The concept of uniformitarianism is a state of the belief that the processes that form and change rocks on and within Earth today are the same processes that have been at work throughout the long geologic history of Earth.
2. It is believed that sufficient pressure and heat exist at the interface between the lithospheric plates and the asthenosphere so that the asthenosphere remains plastic enough to slowly adjust to changes in the distribution of mass above it. This means that the lighter continents and ocean basins tend to "float" on the denser semifluid rock of the asthenosphere.
3. Rapidly decreasing glacial ice reduced the weight of overburden on the continental plates quite rapidly and very recently, in terms of the geologic time required for most major changes in Earth's crustal structure. Isostatic adjustments to compensate for redistribution of mass by erosion and deposition are much slower and occurred earlier in geologic time, making the records much less clear and harder to interpret.
4. The area around the Baltic Sea was covered with massive ice flows that caused the underlying crustal rock to sink into the mantle. Now that the ice has melted, the Baltic Sea region is rising at a rate of nearly 1 meter per century to reestablish the isostatic balance. It appears that this rise will continue until most of this area has risen to a height above sea level, which would cause most of the water to empty from the Baltic Sea.
5. Large horizontal forces due to the movement of lithospheric plates are continually compressing rock, but rock can compress only a limited amount before it begins to buckle and fold.
6. Thrust faulting is a special case of reverse faulting in which the fault plane is at a small angle to the horizontal. This causes one of the plates to slide over the other, descending plate. In normal faulting the overlying side moves downward because the primary factors involved are expansion forces, instead of the compression forces that cause thrust faulting.
7. Mountain systems are made up of smaller geologic units called mountain ranges. Mountain systems are collectively referred to as mountain chains when they occur as elongated units that cover extensive geographic regions.

8. Volcanic mountains are caused by the buildup of lava, cinders, and other pyroclastic debris from the eruption of volcanoes, usually along the edges of subduction zones where one plate is diving under another, producing melting and outgassing of the subducted lithospheric material. Exceptions occur when volcanic activity occurs over hot spots in the mantle, such as the one currently responsible for the volcanic activity in the Hawaiian Islands.
9. The Alps and the Himalayas are both examples of folded mountains that were formed by the collision of two giant continental plates that crashed together during their movement across Earth's surface as the result of plate tectonic activity. The upthrusting caused by the tremendous forces experienced in such a collision has produced some of the highest mountains on Earth.
10. A depositional basin is a long, narrow, subsiding ocean trough containing large accumulations of sediment. The accumulation of large quantities of sediment produces additional weight within the basin that forces this area of seafloor even deeper into the mantle, thus allowing for even more accumulation of sedimentary burden in this area and a related shift in the isostatic balance of Earth's crust.

Chapter 23

Geologic Time

Organizing the geologic history of Earth into a number of time divisions is a useful method geologists have developed during the past several hundred years. The method, for the most part, is based on the worldwide differences noted in living organisms whose remains have been preserved as fossils.

The principle of faunal succession states that certain groups of plants and animals live only during a given time period and are then succeeded by another group in a definite and determinable order. Because the organisms lived during only one period of geologic time, the sediments that now contain their preserved fossils must have been deposited at the same geologic time. This fact provided geologists with a method for developing a relative geologic time scale.

The discovery of radionuclides in rocks and the refinement of techniques for determining the absolute age of rocks in years, rather than just their ages relative to one another, has been achieved in the past 70 years or so. These absolute dating procedures have verified the relative geologic time scale, and both methods are used today in the continued search to decipher Earth's geologic history.

This chapter discusses a chronological history of Earth in both relative and absolute time.

DEMONSTRATIONS

Fossil sets can be used to demonstrate the major types of fossils, their identification, and the geological time periods they represent. Earth history kits, models, and paleoplaques are available from scientific equipment suppliers.

ANSWERS TO REVIEW QUESTIONS

1. b
2. d
3. Relative geologic time is arranging, in the correct sequence, the events that have occurred in the history of rocks.
4. The law of superposition is the simple observation that in a succession of stratified deposits, the younger layers lie on top of the older ones.
5. The relative geologic time scale is based on the characteristic fossils of each stratum.
6. Correlation is determining whether the separate rock sequences for different areas were formed at the same time or at different times.

7. A fossil is a remnant or trace of an organism preserved from a past geological age, such as a skeleton, footprint, or leaf imprint embedded in Earth's crust.
8. Unconformities are breaks or local unrecorded intervals in the geological record.
9. c
10. b
11. Relative geologic dating is based on characteristic fossils found in sedimentary rocks. Absolute geologic dating is based on the radioactive decay of certain atomic nuclei.
12. Half-life is the time required for half of the parent element to decay.
13. Carbon-14 has a relatively short half-life and can be used only for comparatively young specimens.
14. Argon-40, the daughter nuclide of potassium-40, may have escaped from the rock specimen and may give inaccurate results in determining the age of the specimen.
15. e
16. b
17. Cenozoic means recent life.
    Mesozoic means intermediate life.
    Paleozoic means ancient life.
    The Precambrian Era is the earliest era of Earth's history and is so named because the earliest period of the Mesozoic Era, which follows the Precambrian Era, is named the Cambrian Period.
18. Cenozoic—65 My to present. (My stands for million years.)
    Mesozoic—225 My to 65 My.
    Paleozoic—570 My to 225 My.
    Precambrian begins at 4600 My ago and continues to 570 My.
19. Precambrian; about 87%.
20. Cambrian.
21. Silurian.
22. The components of sedimentary rock may not be the same age as the rock layer in which they are found. The sediments that form the rock have been weathered from rocks of different ages.

ANSWERS TO CRITICAL THINKING QUESTIONS

1. (a) The law of superposition.
   (b) The fossil record in rocks.
   (c) Cross-cutting of igneous rock through other rocks.
   (d) Inclusions—pieces of one rock unit that are found in another.
2. Fossils represent life of former times, and the flow of evolution can be deduced from them. The sequence of events can be determined by matching the evolutionary changes revealed by fossils in the rock strata.
3. Widespread extinctions brought on by massive degradation of large ecosystems occur regularly in the fossil record. A vacant ecological space follows, and this is when evolution becomes most creative. A variety of new life begins, replacing that which has become extinct. Thus, the changes brought about by extinctions influence greatly the history of life.

4. (a) The oldest Moon rocks.
   (b) The age of stony meteorites.
   (c) Extrapolating back from the oldest rocks on Earth.

ANSWERS TO RELEVANCE QUESTIONS

23.1. The answer is dependent on where the students are located. Are they finding fossil plants, animals, clam shells, or something else that indicates past environmental conditions?

23.2 No. The most accurate dating is from igneous rock. Grains of a sedimentary rock are not the same age as the rock in which they occur; therefore, the age of the minerals tells us only that the rock can be no older.

23.3 Explain the role of fossils in dating rocks. Take the person to a road cut that displays sedimentary rock layers and show him or her fossils. Briefly explain radioactive dating.

ANSWERS TO STUDY GUIDE QUIZ

Multiple-Choice Questions

1. c	2. b	3. d	4. c	5. a
6. a	7. b	8. d	9. b	10. d

Short-Answer Questions

1. In stratified sedimentary rock, the top layers must have been deposited more recently and so must be younger than the lower layers.
2. The principle of cross-cutting states that any rock layer or fault that cuts across another rock layer or fault must be younger than the original material through which the cut was made.
3. An unconformity is a break in a geologic record that shows that depositing of sedimentary layers stopped for a period of time and erosion began. For this concept to be really useful, the sedimentary depositing process must resume, so that the unconformity represents a well-defined period of erosion. These breaks in the sedimentary record are quite useful because they tend to mark a decided change in climate or geologic activity in that area.
4. Fossils are the remnants of organisms preserved from some past geologic time. If a pattern of change can be established for such organisms and the same fossils are found in widespread areas of Earth's crust, these specimens can be used to correlate the age of the various rock formations in which the fossils are found. Fossils that meet these criteria are known as index fossils.
5. The cooling process of Earth's core was not as straightforward as Lord Kelvin thought, because radioactive material within the Earth's interior provides a continuing source of heat even today.
6. The three most common types of radioactive decay are alpha decay, beta decay, and gamma decay.
7. Radioactive materials do not decay all at once, but proceed through their radioactive decay process in a way that allows only half of the sample to decay in a certain period of time. The same amount of time is required for half of the remaining sample to decay, and so forth. The time for one-half of the sample to decay is known as the half-life of the radioactive material. If the amount of parent nuclei and the amount of daughter nuclei can be accurately determined, the ratio of these amounts can be used as an indication of the age of the rock sample.

8. Potassium-40 decays in a branching mode, with the most abundant daughter being calcium-40. Since calcium-40 is quite prevalent in most rock samples, this branch of decay is very unreliable. The other branch, which produces argon-40 in much smaller quantities, is more useful in providing radioactive dating results.
9. Successful radioactive dating requires that:
   (1) There has been no addition or subtraction of either parent or daughter nuclei in the rock sample other than that caused by the radioactive decay process itself.
   (2) No trace of daughter element was present in the rock sample when it was originally formed; or if there was, the amount can be determined accurately.
   (3) The age of the rock under study differs by no more than a factor of 10 from the half-life of the radionuclide used in the dating process.
10. Sedimentary rock is made up of previously formed and then eroded rock material that could have contained either parent or daughter nuclei before the new sedimentary rock was formed. This makes using radioactive dating processes quite difficult for sedimentary rock samples and puts the results of any such dating in question.

Chapter 24

Surface Processes

Some internal processes, such as volcanism and mountain building, build up Earth's surface features. Many surface processes work in the opposite direction, wearing away and leveling surface features. The weathering and erosion of surface materials are important geologic processes, as is mass wasting, which is the downward movement of weathered materials under the influence of gravity. Evidence of these processes is commonplace. However, some processes take place very slowly and are often overlooked.

Erosion is carried out by the so-called agents of erosion—running water, ice, wind, and waves—which are the physical phenomena that supply the energy for the removal and transportation of rock debris. Two of these agents, running water (streams and rivers) and ice (glaciers), will be examined in some detail.

Not only is water an important factor in geologic processes, it is also necessary to sustain life and is a major environmental concern. In this chapter, Earth's water supply is studied. This supply is often thought to be inexhaustible, because about 71% of Earth's surface is covered by water. However, only about 2% of this is fresh water, and most of it is locked in glacial ice sheets. Earth's water supply is a reusable resource that is constantly being redistributed. In general, atmospheric processes move moisture from large oceanic reservoirs. The water eventually flows back to the seas, eroding as it goes.

Groundwater is an important aspect of this gigantic hydrologic cycle, particularly for domestic water supplies. The geologic features of wells and springs, as well as the quality of water with respect to dissolved minerals, are considered. Also, geothermal hot springs are discussed in the context of an energy resource.

Because oceans cover about 71% of Earth's surface, they are important factors in surface processes. The movement of sea water affects regional climates and is an agent in the erosion of coastlines. The final topic of the chapter deals with the topology of the seafloor. Recent investigations have greatly expanded our knowledge of the surface features of this major unseen portion of Earth's surface.

DEMONSTRATIONS

Geology models show geological land forms in three dimensions. Stream tables use flowing water and rainfall to show the effects of water on land-formation processes such as erosion and sedimentation. Topographic relief maps can be used to show landform features, and contour models illustrate the fundamentals of contour mapping. These models are also very useful in the laboratory when the student is asked to draw contour maps.

ANSWERS TO REVIEW QUESTIONS

1. d
2. e
3. Weathering is the physical disintegration and chemical decomposition of rock.
4. Physical weathering involves the mechanical or physical disintegration of rock, and chemical weathering involves chemical processes that result in chemical changes in rock composition.
5. Frost wedging is the disintegration of rock resulting from the pressure of freezing water. Frost heaving is the lifting of material because of the expansive pressure of freezing water.
6. "Salt wedging," or the disintegration of rock due to the growth of salt crystals.
7. Burrowing animals loosen soil and bring it to the surface. Plant root systems fracture rocks.
8. Primarily moisture, temperature, and the mineral content of the rock.
9. See test material, Section 24.1, for description.
10. A depression resulting from the collapse of a cavern that has been formed by chemical weathering.
11. d
12. b
13. Erosion is the downslope movement of surface and near-surface materials as a result of gravity and the action of agents that cause such movements.
14. Running water, ice, wind, and waves.
15. Erosion resulting from the overland flow of water.
16. Dissolved, suspended, and bed loads. See Section 24.2 for explanation.
17. Traction is the movement of a stream's bed load by current action, which results in the fracture of rocks due to abrasive contact.
18. A mixture of fine soil particles and organic material.
19. Youth, maturity, and old age. See Section 24.2 for a description of each.
20. Meandering results from accumulated deposits of sediment that divert a river's path. A river is graded when its erosion and transport capabilities are in balance.
21. Continental glacial ice sheets cover large areas and flow outward. Valley glaciers form in valleys and flow down in the valleys.
22. Yes. Alaska has valley glaciers, and cirque glaciers occur in the western conterminous United States.
23. (a) Any type of glacial sediment deposit—i.e., from ice or melt water.
    (b) Material deposited by ice rather than melt water.
    (c) A ridge of till.
24. The downward movement of overburden under the influence of gravity.
25. Fast mass wasting: (a), (c), and (e). Slow mass wasting: (b) and (d). See text material, Section 24.2, for process descriptions.
26. c
27. c
28. About 2%, and the majority is frozen in glacial ice sheets.
29. The cyclic motion of Earth's water supply is explained in Figure 24.7.

30. (a) A measure of a material's capacity to transmit fluids.
    (b) The percentage volume of unoccupied space in the total volume of a substance.
    (c) The upper soil region that contains chiefly air between the particles.
    (d) The region below the zone of aeration that is saturated with water.
    (e) The boundary between the zones of aeration and saturation.
31. A body of permeable rock through which groundwater moves.
32. Gravity springs result from the gravitational flow of water. Artesian wells and springs are characterized by water spurting and bubbling to the surface as a result of the special geometry of impermeable rock layers.
33. Geothermal heating below the surface.
34. Surface springs and deep reservoirs. The latter are more important in electrical generation.
35. Dissolved minerals, which can affect soap detergency and can precipitate as scale deposits—e.g., in boilers.
36. b
37. d
38. c
39. A measure of the saltiness of water, expressed in ppt or g/kg. The oceans have an average salinity of 35 ppt or 3.5%.
40. Surface waves, long-shore currents, and tidal currents.
41. In shallow water, the water particles cannot follow circular patterns in the bottom parts of their paths, and the wave crest falls forward.
42. Prevailing surface winds. See Figure 24.23.
43. There are fewer landforms in the Southern Hemisphere to break up circulations. The West Wind Drift is a global circulating current.
44. By warm water carried by the North Equatorial Current, the Florida Current, and the North Atlantic Current.
45. Currents caused by dense water sinking through less dense water. Deep-water currents transport water toward the equator in ocean circulations.
46. A shallow barrier in the Bering Strait prevents cold Arctic water from flowing southward.
47. (a) Isolated submarine mountains.
    (b) Flat-topped seamounts.
    (c) Trenches that mark plunging plate boundaries.
    (d) Large flat areas of sediment deposits on the ocean floor that cover the original topology.
48. Continental shelves are the relatively shallow submerged borders of the continents. Continental slopes are the true edges of the continents, where the continental land masses slope downward to the floors of the ocean basins.
49. Because of the fishing rights and offshore oil drilling.

ANSWERS TO CRITICAL THINKING QUESTIONS

1. (a) Modification of the slope geometry.
   (b) Reduce the load—remove part of the slope mass.
   (c) Decrease the water content.
   (d) Plant vegetation with sturdy roots.
   (e) Construct retaining walls.
2. (a) Contour plowing.
   (b) Avoid plowing on steep slopes.
   (c) Plant vegetation with sturdy roots.
   (d) Proper drainage of water.
3. Dust the surface of the glacier with a dark material to increase the absorption of the Sun's rays. This will increase heating of the surface, which will increase the rate of melting and the water flow.
4. The answer depends on the geographic location of the student.

ANSWERS TO RELEVANCE QUESTIONS

24.1. When rocks are exposed to air, water, and organic matter, disintegration (mechanical breakdown) or decomposition (chemical decay) takes place. Student answers will vary, including weathering done by frost wedging, acid rain, wind, oxidation, leaching, and hydration.

24.2 The answer will depend on the student's location. At a lake or ocean, shoreline erosion will be by water waves. At a river or stream, erosion will be by running water. Erosion may involve the movement of sediments by wind (dust storms) or wind abrasion. Man-made erosion is caused by off-road recreational vehicles.

24.3 The local primary water source will most likely be a river, lake, well, or spring.

24.4 Answers will depend on the student's location, but the comparisons are as follows:
Seamounts—land mountains
Guyots—flat-topped mountains
Mid-ocean ridge—mountain range
Seafloor trenches—deep valleys
Abyssal plains—prairies

ANSWERS TO STUDY GUIDE QUIZ

Multiple-Choice Questions

1. b	2. a	3. c	4. a	5. c
6. c	7. b	8. d	9. a	10. d

Short-Answer Questions

1. Heat and moisture are the two most important factors in determining the rate of chemical weathering.
2. In cold regions, the subsurface soil may remain frozen permanently, creating a permafrost layer. This can cause the topsoil to become wet and spongy during the few weeks in the summer when the topmost foot or so thaws. This occurs because the permafrost provides a stable base that prevents the melted water from draining.

3. Three types of stream loading are described below.
   (1) Dissolved load occurs when water-soluble minerals are carried along in solution by the stream.
   (2) Suspended load is made up of fine particles that are not heavy enough to sink to the bottom and so are carried downstream by the moving water.
   (3) Bed load is composed of coarse particles and rocks that are rolled or bounced along by the current near the bottom of the stream in a process called traction.
4. Abrasion occurs when the coarser load material near the bottom of streams is broken into smaller rocks and particles, producing smooth, rounded rocks and pebbles. This occurs most often in streams that have swift currents and large volumes of flow.
5. Although Earth's overall water supply is quite large, about 98% of that water is in oceans, where the salt content makes it unfit for human consumption without putting it through a costly desalination process. Most of the remaining 2% is fresh water, but a large portion of it is frozen in the glacial ice sheets of Greenland and Antarctica. This leaves a very limited supply of fresh water available for use by humans.
6. An old river is characterized by (1) meanders (looping twists and turns in the river's path), (2) grading (little deepening of its channel because erosion is balanced by transport depositing), and (3) side flood plains.
7. A gravity spring is simply a flow of groundwater, caused by gravity, that emerges naturally onto Earth's surface at or below the surrounding groundwater level. In an artesian spring, the flowing underground water is trapped between impermeable layers of rock, which causes it to build up pressure. This can produce a discharge of water above the level of the surrounding groundwater.
8. Three types of landslides are
   (1) Rockslides—large quantities of rock break off and move rapidly down the steep slopes in mountainous regions.
   (2) Slumps—slow downslope movement of unbroken blocks of overburden, resulting in a curved depression above the slump on the slope.
   (3) Mudflows—loose ash or other material, often on volcanic slopes, absorbs water from local, heavy rainstorms and then flows downslope, often with disastrous results.
9. A long-shore current arises from the action of incoming waves that break on the shore at an angle. The component of water motion parallel to the shore causes a directional current that moves both water and debris along the shoreline.
10. Undersea volcanic mountains are often found as individual, isolated seamounts. When these have flat tops, they are known as guyots. The flattened tops of these seamounts suggests that they were once islands at the ocean surface, where erosion by wave action removed the rock that projected above sea level. Subsequent subsidence caused these islands to sink below sea level, sometimes to depths of several hundred feet, where they are found today with their flattened tops. No underwater erosion process that can produce such flattened tops is known, so the subsided-island theory seems to be the best explanation.

Chapter 25

The Atmosphere

For students to understand and appreciate weather principles and phenomena, it is essential that they know the fundamental physical properties of the atmosphere. The current emphasis on the environment offers a timely introduction to the study of the atmosphere. Everyone is aware of atmospheric pollution problems (Chapter 26); however, to be really knowledgeable, students should know the normal conditions of our atmosphere that are being jeopardized. These conditions are presented in this chapter.

The chapter discusses the composition of the atmosphere and how the relative percentages of the major constituents are maintained. Attention is given to the physical properties that distinguish one part of the atmosphere from another, particularly the divisions based on temperature. The section on the energy content of the atmosphere gives the student an understanding of what is responsible for the dynamics of the atmosphere. As pointed out in Chapter 1, measurement is necessary to describe conditions, and this also applies to the atmosphere. Common atmospheric measurements, with which the student should be familiar, are discussed. The concept of relative humidity and the dew point temperature should be covered thoroughly.

The last two sections of the chapter are concerned with the movements of the gases of the atmosphere—winds and air currents—and the associated formation of clouds. Convection cycles and their relationship to local winds and world circulation patterns have important effects to which the student can relate. This is what makes the study of meteorology so interesting.

Clouds are one of the most common sights in a student's environment. This promotes interest in the study of various cloud types and characteristics. A knowledge of cloud formation is prerequisite to understanding precipitation processes, mechanisms, and types, which will be considered in the next chapter.

DEMONSTRATIONS

1. Atmospheric pressure

A dramatic demonstration of atmospheric pressure is the crushing of a metal can. There are two variations of this demonstration. One way is to use a gallon metal can (e.g., a *well-rinsed* paint thinner can). Place a small amount of water in the can and heat the *open* can sitting on a ring stand with a Bunsen burner. When the water is boiling vigorously, as evidenced by condensed water vapor coming from the can (not steam, which is invisible), ask the students to *turn off the burner* and place a rubber stopper securely in the can opening. (A rubber stopper is better than the screw, gasketed cap. The gasket may leak, and the rubber stopper is safer should you forget to turn off the burner.) As the can cools and the steam condenses, a partial vacuum forms and the pressure difference causes the can to slowly collapse. After the can is crushed, slowly remove the stopper. An audible rush of air illustrates the partial vacuum in the can (like that in a vacuum-packed coffee can).

A quicker, and perhaps more convenient, way to demonstrate atmospheric pressure is to use an aluminum soft drink can. (Large metal cans are sometimes difficult to obtain.) Place a small amount of water in the rinsed can, and hold the can over the flame of a Bunsen burner using a pair of tongs. Upon hearing vigorous boiling, quickly invert the top of the can into a container of water. The can is crushed immediately. When the can is raised from the water, water will run from the can, having been forced into the partial vacuum. (Have an extra can available, because students will want to see the demonstration repeated.)

2. Dew point and atmospheric pressure

For a classroom demonstration of dew point and atmospheric pressure, an iced soft drink in a glass with a straw can be used. After studying the section on humidity, the student should be able to explain why condensation occurs on the glass. This demonstrates that the air in the vicinity of the glass has been cooled to its dew point, and because the air is saturated (100% relative humidity), the water vapor condenses on the glass.

Drinking through a straw demonstrates atmospheric pressure. Aspiration through the straw reduces the pressure within, and liquid is forced up the straw by the pressure difference relative to the outside atmospheric pressure. This can be effectively shown by putting a hole in the side of the straw. This equalizes the pressure and makes the straw useless.

3. Humidity

A simple psychrometer can be easily constructed. The only essential equipment is two thermometers, a water reservoir, and a cloth wick to keep one of the bulbs wet. Care should be taken not to place the wet bulb too near the water reservoir, as this will impair free evaporation. One method is to use a paper quart milk container as a reservoir and attach the thermometers to the side using garbage bag ties. A hole can be made for the wick to extend into the water.

The relative humidity and the dewpoint temperature can be determined from the difference in the thermometer readings by using the tables given in the text Appendix IX. This demonstration can be set up during one class period and a reading taken the next, to allow time for the evaporation rate to come to equilibrium. The students themselves can construct a psychrometer as a project, taking readings in their dormitory for a period of days and plotting the results to show the humidity variation over time.

4. Weather instruments

Weather-measuring instruments, such as barometers and anemometers, can be used to demonstrate how measurements are obtained. See the companies listed in the Teaching Aids section at the end of this *Guide*.

5. Convection cycles

A commercially available product called a Lava-lite works well in demonstrating convection cycles. The Lava-lite consists of a lamp base with an upper portion containing liquids of different densities. One liquid is colored and rises in the other as a result of heat from the lamp. As the globules of liquid rise and cool, they fall back to the bottom of the container. The Lava-lite should be turned on an hour or more before the scheduled demonstration, as time is required to initiate the action.

ANSWERS TO REVIEW QUESTIONS

1. d
2. c
3. In general, atmospheric science deals with all the atmosphere, whereas meteorology was concerned chiefly with weather in the lower atmosphere.
4. 78% nitrogen, 21% oxygen, and 1% other gases.
5. No, photosynthesis by plants replenishes the oxygen.
6. It is essential in photosynthesis.

7. (a) Decreases.
   (b) Increases.
   (c) Decreases.
   (d) Increases.
8. It absorbs and protects us from harmful ultraviolet radiation in sunlight.
9. Solar disturbances, and the recombination of ionized gas molecules and electrons in Earth's upper atmosphere.
10. b
11. c
12. *In*coming *sol*ar radi*ation*.
13. (1) Absorption of terrestrial radiation, (2) latent heat of condensation, (3) conduction from Earth's surface.
14. There is continued heating after the maximum is reached.
15. (a) Rayleigh scattering.
    (b) Least scattering, and so can be seen for the greatest distance.
16. (a) The transmission of visible solar radiation and the absorption of re-radiated infrared radiation by the atmospheric gases in a manner analogous to the glass in a greenhouse.
    (b) A change in temperature shifts the wavelength of the radiation such that it is transmitted or absorbed, and Earth cools and heats periodically.
17. b
18. d
19. The mercury column is supported by atmospheric pressure, 30 in. Hg or 76 cm Hg.
20. An aneroid barometer calibrated inversely in altitude.
21. The temperature of the air in the vicinity of the glass is lowered to the dew point.
22. The evaporation from the wet bulb is inversely proportional to the relative humidity, and the depression of the wet bulb provides a measurement of humidity.
23. In the direction from which the wind is coming, because of the fins.
24. Wind fills the pivoted sock, and the tail of the sock points in the direction toward which the wind is blowing. The angle of the sock relative to the horizontal is an indication of the wind speed. (With a high wind speed, it stands "straight out.")
25. Doppler radar can give wind speed and direction, whereas conventional radar cannot.
26. a
27. b
28. Primary: pressure due to temperature difference and gravity. Secondary: Coriolis force and friction. Secondary forces are velocity-dependent.
29. As viewed from above, in the Northern Hemisphere, cyclones rotate counterclockwise and anticyclones rotate clockwise. The rotations are opposite in the Southern Hemisphere.
30. (a) From west to east, in westerly wind zone.
    (b) West, upwind.
    (c) Windward side may need extra insulation.
31. d
32. b

33. (a) Low.
    (b) High.
    (c) Middle.
    (d) Low.
    (e) Vertical development.
34. (a) Cirrocumulus.
    (b) Cirrostratus.
    (c) Stratus.
    (d) Cumulonimbus.
35. Cooling air to dew point (e.g., rising air) and hygroscopic nuclei.
36. The air stops rising and no cloud is formed.

ANSWERS TO CRITICAL THINKING QUESTIONS

1. They have small masses and gravity is not sufficient to retain energetic gas molecules of atmosphere. They are relatively close to the Sun, and so the atmospheric temperature is high.
2. About 5 times (0.33/0.07 = 4.7).
3. (a) Lower specific heat, less heat to lose. Also, not mixed as in water.
   (b) Very dry and little water vapor or clouds to absorb terrestrial radiation and insulate from heat loss.
4. 100% relative humidity.
5. Clockwise and counterclockwise are "senses," like right and left. Direction depends on orientation or reference.
6. Rising air at the equator, setting up a convection cycle with surface winds toward equator (from north to south in Northern Hemisphere and south to north in Southern Hemisphere).

ANSWERS TO EXERCISES

1. (a) (50 – 16 km)/16 km = <u>2.1</u>
   (b) (80 – 50 km)/16 km = <u>1.9</u>
   (c) (200 – 80 km)/16 km = <u>7.5</u>
2. Graph.
3. $T = T_0 - Rh$ = 70°F – (3.5 F°/1000 ft)(14,000 ft) = <u>21°F</u>
4. $T = T_0 - Rh$ = 20° – (6.5 C°/km)(10 km) = <u>–45°</u>
5. From tables:
   (a) 70%
   (b) 64°F
   (c) 9.4 gr/ft^3
   (d) $AC = MC \times RH$ = (9.4 gr/ft^3)(0.70) = <u>6.6 gr/ft^3</u>

6. From tables:
   (a) 82%
   (b) 89°F
   (c) 17.1 gr/ft^3
   (d) AC = (17.1 gr/ft^3)(0.82) = 14 gr/ft^3

7. (a) $AC = MC \times RH$ = (23.4 gr/ft^3)(0.90) = 21 gr/ft^3
   (b) DP = 101°F (table), and 105°F − 101°F = 4 F°

8. (a) AC = (2.4 gr/ft^3)(0.45) = 1.1 gr/ft^3
   (b) No. Coolness is due to water evaporating (latent heat removed from bulb). Air temperature is above freezing, and water is replaced from reservoir (also at 35°F).

9. (a) $AC = MC \times RH$ = (9.4 gr/ft^3)(0.58) = 5.5 gr/ft^3
   (b) DP = 59°F (from table)

10. ΔT = 5 F° (from table) and DP = 73°F (from table), so 80°F − 73°F = 7 F°

ANSWERS TO RELEVANCE QUESTIONS

25.1 The ultraviolet radiation passes through the clouds, whereas the infrared radiation (heat rays) does not. Hence, on a cloud-covered day, we don't feel as hot, but the burning and tanning UV radiation is still received.

25.3 All of them—temperature, pressure (barometer reading), humidity, wind speed and direction, and precipitation. Satellite and radar observations are also commonly seen.

25.4 In the conterminous United States, generally from the west (Westerlies wind zone), although there may be other influences if you live near large bodies of water or near the boundaries of the wind zone.

25.5 Depends on location, but now you know the cloud families and types. If there are no clouds, then your region is probably experiencing a "high," and there is no rising air for cloud formation.

ANSWERS TO STUDY GUIDE QUIZ

Multiple-Choice Questions

1. d	2. d	3. a	4. c	5. b
6. b	7. b	8. c	9. a	10. d

Short-Answer Questions

1. Animals, including humans, breath in oxygen and exhale carbon dioxide while plants use carbon dioxide in the photosynthesis process and release oxygen as a by-product. These processes tend to balance each other and leave the oxygen/carbon dioxide levels in the atmosphere nearly constant.

2. From Earth's surface upward, the main vertical divisions of the atmosphere are: the troposphere, the stratosphere, the mesosphere, and the thermosphere.

3. Ozone is a three-atom molecule of oxygen (O_3) that can be found throughout the stratosphere. Ozone is formed by the interaction of diatomic oxygen with energetic ultraviolet light and, once formed, acts as an umbrella that shields life forms on Earth from harmful short wavelength ultraviolet radiation produced by the Sun.

4. The speed of the wind is measured by an instrument known as an anemometer. Three or four cups attached to a rotating shaft, much like a pinwheel, catch the wind so that the rotation of the shaft serves as an indication of the wind speed. This device can be combined with a wind vane to indicate the direction, as well as the speed of the wind.

5. The greenhouse effect is a process that traps heat from the Sun in our atmosphere much like heat is trapped in a glass covered "greenhouse" used to warm and protect plants in the early Spring and late Fall. The atmosphere allows visible light to pass through quite freely to the surface of Earth where it is absorbed by the ground and other dark-colored surfaces. This absorbed energy is reradiated into the lower atmosphere as infrared radiation, which is much more easily absorbed by the surrounding air, thus trapping the heat from the Sun.

6. Gravity and the pressure differences due to temperature variations provide the two primary forces that affect the motion of the air surrounding our Earth; that is, they produce winds.

7. Heated air tends to rise and when it does, cool air flows in from the sides to fill in the vacated space. During the day, the land is heated by the Sun more than the adjacent water (ocean or lake) and so the wind tends to blow in from the area over the cool water toward the land and a sea breeze is produced. At night the land cools off more rapidly, leaving the water warmer. This means the rising air is now over the water and wind from the shore sweeps out to fill in the vacated area. This results in a land, or off-shore, breeze after the Sun goes down.

8. Large moving rivers of air, high up in the troposphere, called jet streams influence the movement of air masses across the surface of Earth and thus affect and sometimes even dominate the weather patterns produced by these air mass movements.

9. Radar may be used to detect and monitor precipitation and is, therefore, useful in determining the location of severe storms. Conventional radar installations monitor weather all across the United States and are also found in many other parts of the world, especially near airports and large cities. Doppler radar, or more properly Doppler shift radar, can not only detect the distribution and intensity of precipitation, but can also determine the horizontal motion of the rain or snow and thus map out the direction and speed of the associated winds that carry this precipitation.

10. Clouds begin to form when rising, moist air reaches its dew point and the water vapor in the air begins to condense out, forming large droplets that make up the cloud itself. The cloud continues to be formed as the moist air rises until the temperature of the rising air and the surrounding air mass become the same, at which point the moist air ceases to rise and the top of the cloud formation is reached.

Chapter 26

Atmospheric Effects

Having covered the fundamentals in the preceding chapter, we now turn our attention to the processes and dynamics of weather phenomena. Since we have studied cloud formation, we can now look at precipitation processes and types. This includes a discussion of the effect of bacteria on the formation of ice nuclei and frost.

In general, the movement of large air masses across the country affects our weather changes. A person who knows about the movements and characteristics of these air masses is better prepared to understand and predict changes in the weather. In this chapter, students will encounter terms familiar to them from weather forecasts, such as *front* and *polar air masses*. The material is easily related to daily observations, and particularly at this point, students should be encouraged to understand how meteorological conditions affect the environment.

A section is devoted to sensational weather phenomena—storms. These atmospheric disturbances sometimes give rise to property damage and even death. The properties and characteristics of some of the most powerful storms are discussed. Also, safety procedures are emphasized, particularly for tornadoes and hurricanes.

The environmental and climatic effects of air pollution are major concerns today. Discussions of these effects commonly appear in daily newspapers and can be heard on radio and TV. Having studied the atmosphere, the student can now better understand and appreciate these problems. A brief history of air pollution is given, and the chief pollutants are identified, along with their sources. Discussions such as the one on acid rain point out how pollution affects our environment and our lives. Also, the ramifications of ozone depletion and the ozone hole over the South Pole are discussed. The long-term climatic effects of atmospheric pollution, for which we are responsible, can only be speculative.

DEMONSTRATIONS

Chemical reactions of some air pollutants—SO_2, for example—can be demonstrated in the classroom. However, the utmost care should be taken, and this type of demonstration is not advised. A safer demonstration is the collection of particulate matter by using filter paper. A damp filter paper can be left exposed in the classroom, or large volumes of air can be filtered for particulates by using a vacuum cleaner. The particulate matter on the filter is then examined with a scanning microscope. Film slides may be made if such equipment is available. As a class project, students can collect samples at different outside locations. Commercial air-sampling kits are also commercially available.

ANSWERS TO REVIEW QUESTIONS

1. b
2. a
3. Bergeron process essentials: (1) ice crystals, (2) supercooled water vapor, and (3) mixing. Silver iodide crystals are substituted for ice crystals, and dry ice is used to cool vapor and form ice crystals.
4. (a) No. Frost is the deposition of water vapor.
   (b) Successive cycles and condensation in cumulus clouds.
5. a
6. d
7. According to the surface and the latitude of their source regions. If the source region is a water surface and at low latitude, the air mass would be warm and moist. Similarly, land areas at high latitudes provide cold and dry air-mass characteristics.
8. (a) See Table 26.1 and Fig. 26.3.
   (b) Alaska: cP, mA, and mP. Hawaii: mT, and possibly mild mP.
9. The boundary of air masses.

   Warm ⌓⌓⌓⌓,

   cold △△△△,

   stationary △◡△◡, and

   occluded △⌓△⌓.
10. See Section 26.2 for descriptions. The sharpness of a front's vertical boundary gives an indication of the rate of change of the weather. In general, cold fronts have sharper vertical boundaries than warm fronts, and hence lead to more sudden weather changes.
11. b
12. d
13. Because of updrafts associated with the low pressure of the storm center. It is believed that a charge separation occurs as water droplets are broken apart.
14. Resuscitate and keep warm. (Have someone call 911, or you do so as soon as possible after resuscitating.)
15. Warm front. A warm front advances over colder air. If the temperature of the cold air and Earth's surface is below freezing, precipitation falling may cool and freeze on contact, producing an ice storm.
16. The tornado. Although the hurricane has more energy, the energy of a tornado is concentrated in a small region, giving a greater energy density. The larger hurricane, however, is usually more destructive on making landfall.
17. Updrafts produce a funnel vortex that is outlined by clouds and debris. Winds are from 300 mi/h, and the tornado is unpredictable in direction and relatively short-lived.
18. Latent heat. When the wind speed reaches 74 mi/h (118 km/h).
19. A hurricane watch is issued for coastal areas when there is a threat of a hurricane within 24 to 36 hours. A hurricane warning indicates that hurricane conditions are expected within 24 hours.

20. (a) August–September, for the North Atlantic.
    (b) Varies from state to state. Generally, April–August.
21. a
22. b
23. Any atypical contribution to the atmosphere resulting from human activities.
24. No. England had air pollution in the late 1200s.
25. Combustion (incomplete) and fossil fuel impurities.
26. Complete—CO_2 and H_2O; incomplete—CO, hydrocarbons, and soot (carbon).
27. Complete. The high temperatures of complete combustion cause a reaction between the nitrogen and oxygen of the air.
28. Classical smog is smoke-fog. Photochemical smog results from photochemical reactions of hydrocarbons and other pollutants with oxygen in the presence of sunlight. Ozone is the prime indicator of photochemical smog.
29. Sulfur.
30. Sulfur dioxide (and nitrogen oxides) combine with water to form acids, which fall as rain (or other types of precipitation). The acid raises the pH of bodies of water. The problem is most acute in the northeastern United States because of major industrial areas to the west, but acid precipitation may now be found almost everywhere.
31. (a) Transportation.
    (b) Stationary sources (e.g., electrical generating plants).
    (c) Industry.
    (d) Transportation.
    (e) Photochemical smog.
32. d
33. c
34. From 1880 to 1940, the temperature increased by about 0.6 C°/y. Since 1940, the temperature has decreased by about 0.3 C°/y.
35. Increased concentrations of CO_2 in the atmosphere could alter the amount of radiation absorbed and give rise to an increase in Earth's temperature.
36. Increased CO_2, increased particulate matter, and ozone reduction.
37. The interaction of CFCs with the ozone layer could deplete the ozone and allow more UV to reach Earth, thereby increasing Earth's temperature. In the extreme, an increase in the average temperature could affect the environment (e.g., lengthen the growing season) and melt the polar ice caps.
38. There are no natural mechanisms in the stratosphere to remove pollutants, and the pollutants could give rise to changes in climate.

ANSWERS TO CRITICAL THINKING QUESTIONS

1. In general, fair weather is associated with high pressure, since the pressure reduces the formation of clouds and hence precipitation. Bad weather is associated with low pressure, which allows cloud formation and precipitation.
2. Yes. Alternative energy sources that do not produce CO_2, such as nuclear generating stations and solar energy.

ANSWERS TO EXERCISES

1. See Fig. 26.3.
2. (a) cT
   (b) mA
   (c) cA
   (d) mT
   (e) cP and cA.
3. $d_c = v_c t = (35\text{ km/h})(24\text{ h}) = 840\text{ km}$
   $d_w = v_w t = (20\text{ km/h})(24\text{ h}) = 480\text{ km}$
   $\Delta = 840\text{ km} - 480\text{ km} = \underline{360\text{ km}}$
4. Personal. $t = d/v$, where average v from Exercise 3 may be used.
5. $d = vt = (1/3\text{ km/s})(4.0\text{ s}) = \underline{1.3\text{ km}}$
6. $d = vt = (1/5\text{ mi/s})(11\text{ s}) = \underline{3.7\text{ mi}}$

ANSWERS TO RELEVANCE QUESTIONS

26.1 It evaporates when the air temperature rises above the dew point.

26.2 Answer depends on location.

26.3 Watch: Be alert for possible tornado formation.
Warning: Follow tornado safety procedures given in the chapter.

26.4 Everyone contributes either directly or indirectly (for example, through the use of manufactured products and overconsumption).

26.5 No, climatic changes involve *long*-term averages. Annual variations could be due to a variety of things—for example, a seasonal shift in the jet stream patterns.

ANSWERS TO STUDY GUIDE QUIZ

Multiple-Choice Questions

1. a	2. d	3. b	4. a	5. c
6. b	7. c	8. a	9. c	10. b

Short-Answer Questions

1. Since cT stands for continental-tropical, we can expect this air mass to be relatively dry because it formed over land, and warm because it formed near the equator.
2. When a cold front enters an area, the temperature will drop and the approaching cool, dense air will push under the local warmer air, forcing it to a higher altitude, where rain will quite often be formed. Cold fronts have sharp vertical boundaries that displace warm air quite rapidly, thus producing violent and sudden storm conditions.
3. When a thunderstorm is producing lightning in your area, you should follow the following safety rules:
   (1) Stay indoors away from open windows, fireplaces, and electrical appliances such as refrigerators, stoves, or sinks.
   (2) Do not use the telephone.
   (3) Do not use plugged-in electrical equipment such as lamps or radios.
   (4) If caught outside, seek shelter in a building or in a ditch or ravine. If your hair stands on end or your skin tingles, drop to the ground immediately.
4. During an ice storm, the ground temperature is below freezing; however, falling rain does not freeze until it strikes a cold surface. This means that ice builds up on any cold surfaces that are exposed to the rain, often causing severe damage to trees, power lines, and even buildings.
5. Wind speeds associated with tornadoes are very high, and the energy of the storm is concentrated in relatively small areas, where considerable localized damage is quite likely to occur.
6. A hurricane forms when ascending warm air, heated by the Sun, takes on a spiral motion as a result of the Coriolis effect. This happens over tropical oceanic regions where the air is already warm and moist, so that additional energy is released when the rain condenses out of the rising, high-humidity air. This latent heat is the major source of the great destructive energy found in this type of storm.
7. A hurricane watch is issued for coastal areas that may be hit by the storm within 24 to 36 h. A hurricane warning is given when hurricane conditions are expected to actually occur in the region within the next 24 h.
8. Approaching cirrus and altocumulus clouds indicate the arrival of a warm front in the region. The arrival of a warm front can bring precipitation and storms, but these are generally less severe and more gradual in their onset than those accompanying a cold front. In the case of a warm front, most precipitation occurs before the front has passed.
9. Incomplete combustion of fuels leads to the formation of carbon monoxide, which is a poisonous gas. It also produces soot and other particulate matter that can contribute to air pollution. Even complete combustion of fuels can lead to increased levels of carbon dioxide, which can form carbonic acid, and sulfur and nitrogen oxides, which are also harmful air pollutants.
10. A temperature inversion occurs when the lapse rate in air temperature locally increases instead of decreases with height near Earth's surface. This increase in temperature with increasing altitude keeps hot combustion gases from rising and becoming dispersed over large areas. This can lead to severe localized concentrations of pollution that pose serious health hazards to people living in the area affected by the thermal inversion.

Appendix

Teaching Aids

INTRODUCTION

There are many ways to supplement the physics, chemistry, astronomy, geology, and atmospheric sciences material covered in the textbook. Some of these have been described in the discussion of the individual chapters in this *Instructor's Guide.* The following sections list (1) suppliers of demonstration equipment and supplies, (2) suppliers of audio-visual material, and (3) publications of use to instructors. The lists are, of course, not complete, but they are certainly representative.

SOURCES OF DEMONSTRATION EQUIPMENT AND SUPPLIES

Carolina Physical Science
(Carolina Biological Supply Co.)
2700 York Road
Burlington, NC 27215-3398
1-800-334-5551

Central Scientific Co. (CENCO)
3600 CENCO Parkway
Franklin Park, IL 60131-1364
1-800-262-3626

Edmund Scientific
101 E. Gloucester Pike
Barrington, NJ 08007-1380
1-609-573-6250

Fisher Scientific
711 Forbes Avenue
Pittsburgh, PA 15219-4785
1-800-766-7000

Fisher Scientific
Educational Materials Division
4901 West LeMoyne St.
Chicago, IL 60651
1-800-955-6644

Klinger Education Products Co.
112-19 14th Road
College Point, NY 11356
1-800-522-6252

Metrologic Instruments, Inc. (good for lasers)
P.O. Box 307
Bellmawr, NJ 08099-0307
1-800-436-3876

Pasco Scientific
P.O. Box 619011
Roseville, CA 95611-9011
1-800-772-8700

Sargent-Welch Co.
911 Commerce Court
Buffalo Grove, IL 60089-2362
1-800-727-4368

Scientific Sales, Inc.
P.O. Box 6725
Lawrenceville, NJ 08648
1-800-788-5666

Robert E. White Instruments, Inc. (good for weather instruments)
34 Commercial Wharf
Boston, MA 02110
1-800-992-3045

SUPPLIERS OF AUDIO-VISUAL MATERIAL

AAPT Executive Office
5112 Berwyn Road
College Park, MD 20740
(301) 345-4200

Ambrose Video Publishing, Inc.
28 West 44th Street, Suite 2100
New York, NY 10036
1-800-526-4663

American Geological Institute
4220 King Street
Alexandria, VA 22302-1502
703-379-2480

Astronomical Society of the Pacific
390 Ashton Avenue
San Francisco, CA 94112
1-800-3335-2624

Coronet Film and Video
4350 Equity Dr.
P.O. Box 2649
Columbus, OH 43216
1-800-321-3106

EME Corp
P.O. Box 2805
Danbury, CT 06813-2805
1-800-848-2050

Encyclopedia Britannica Educational Corp.
310 South Michigan Avenue
Chicago, IL 60604
1-800-554-9862

Falcon Software, Inc.
P.O. Box 200
Wentworth, NH 03282
603-764-5788

Films for the Humanities and Sciences
P.O. Box 2053
Princeton, NJ 08543-2053
800-257-5126

Films Incorporated Video
5547 North Ravenswood Avenue
Chicago, IL 60640-1199
1-800-323-4222

Hansen Planetarium
1845 South 300 West "A"
Salt Lake City, UT 84115
1-800-321-2369

Hawkhill Associates, Inc.
P.O. Box 1029
Madison, WI 53701-1029
1-800-422-4295

Indiana University, Media Resources
Bloomington, IN 47405-5901
1-800-552-8620

Insight Media
2162 Broadway
New York, NY 10024
1-800-233-9910

International Film Bureau, Inc.
332 South Michigan Avenue
Chicago, IL 60604-4382
312-427-4545

JLM Visuals
1208 Bridge Street
Grafton, WI 53024
414-377-7775

NASA Johnson Space Center
Media Services Branch/AP3
Film/Video Distribution Library
2101 NASA Road One
Houston, TX 77058
713-483-8696

National Geographic Society
Educational Services
1145 17th Street NW
Washington, DC 20036-4688
1-800-368-2728

National Science Foundation
1800 G. Street NW
Washington, DC 20650

Phoenix/BFA Films and Videos
2349 Chaffee Drive
St. Louis, MO 63146
1-800-221-1274

Sky Publishing Company
P.O. Box 9111
Belmont, MA 02178-9111
1-800-253-0245

The Media Guild
11722 Sorrento Valley Road, Suite E
San Diego, CA 92121
1-800-886-9191

University of California Extension Center for Media and Independent Learning
2000 Center Street, Fourth Floor
Berkeley, CA 94704
510-642-0460

U. S. Department of the Interior (for aerial and space imagery photographs)
EROS Data Center
Geological Survey
Sioux Falls, SD 57198
605-594-6151

U. S. Geological Survey
Photographic Library
MS 914
Box 25046, Federal Center
Denver CO 80225-0046
303-236-1010
(Request: "Geological Survey Photographic Library")

Ward's Multimedia
P.O. Box 92912
Rochester, NY 14692-9012
1-800-962-2660

ZTEC Company
P.O. Box 11768
Lexington, KY 40577-1768
1-800-247-1603

PUBLICATIONS OF USE TO INSTRUCTORS

Recently published textbooks in physics, chemistry, astronomy, geology, and atmospheric science are good reference sources for instructors. In addition, the following magazines and journals are recommended sources of demonstrations and information for teaching physical science.

Astronomy (magazine)
Kalmbach Publishing Co.
P.O. Box 1612
Waukesha, WI 53187
1-800-533-6644

Chemical Demonstrations: A Handbook for Teachers of Chemistry, B. Z. Shakhashiri
Vol. 1 (1983), Vol. 2 (1985), Vol. 3 (1984), Vol. 4 (1992)
University of Wisconsin Press
Madison, WI 53715

Chemical Demonstrations: A Sourcebook for Teachers, Summerlin and Ealy,
OTB series, 1985; Vol. 2, 1987.
(These paperbound books are available from the American Chemical Society.)

Discover (magazine)
P.O. Box 420087
Palm Coast, FL 32142-9944
1-800-829-9132

Earth (magazine)
P.O. Box 1612
Waukesha, WI 53187-9950
414-796-8776

Journal of Chemical Education
1991 Northampton Street
Easton, PA 18042
215-250-7264

Science News (magazine)
P.O. Box 1925
Marion, OH 43305
1-800-247-2160

Scientific American (magazine)
415 Madison Avenue
New York, NY 10017-1111
1-800-333-1199

Sky and Telescope (magazine)
P.O. Box 9111
Belmont, MA 02178-9111
1-800-253-0245

The Demonstration Handbook of Physics, Freier and Anderson
AAPT Executive Office
Publications Department
5112 Berwyn Road
College Park, MD 20740
(301) 345-4200

The Physics Teacher (magazine)
AAPT Executive Office
Publications Department
5112 Berwyn Road
College Park, MD 20740
(301) 345-4200

The Planetary Report (magazine)
The Planetary Society
65 N. Catalina Ave.
Pasadena, CA 91106-2301
818-793-5100

The Universe in the Classroom (a newsletter free to teachers)
Teachers' Newsletter Department
Astronomical Society of the Pacific
390 Ashton Avenue
San Francisco, CA 94112-1787

The Earth in the Classroom (a newsletter free to teachers)
Byrd and Block Communications
719 Patterson Avenue
Austin, TX 78768

Astronomical Society of the Pacific
390 Ashton Avenue
San Francisco, CA 94112-1787

Among the many collections of pictures of clouds are the following:

A Cloud Atlas, McAidie, A. G., Rand McNally, New York
Cloud Types for Observers, British Meteorological Office, Her Majesty's Stationery Office
International Cloud Atlas, World Meteorological Organization, Geneva, Switzerland

Enviromental information may be obtained from:

Office of Public Information
Environmental Protection Agency
401 M St. SW #3404
Washington, DC 20460

"Daily Weather Maps" for the previous week may be obtained from:

Superintendent of Documents
U.S. Government Printing Office
Washington, DC 20401

In addition to the films and videos available from the NASA Johnson Space Center in the audio-visual suppliers list, NASA has educational materials available from the following regional service centers:

(For AL, AR, IA, LA, MS, MO, TN)

NASA Marshall Space Center
Teacher Resource Center
One Tranquility Drive
Huntsville, AL 35807
205-544-5812

(For CT, DE, DC, MA, ME, MD, NH, NJ, NY, PA, RI, VT)

NASA Goddard Space Center
Teacher Resource Laboratory
Mail Code 130.0
Greenbelt, MD 20771
301-286-8570

(For FL, GA, PR, VI)

NASA Kennedy Space Center
Educators Resource Laboratory
Mail Code ERL
Kennedy Space Center, FL 32889
407-867-4090

(For KY, NC, SC, VA, WV)

NASA Langley Research Center
Teacher Resource Center
Mail Stop 146
Hampton, VA 23665
804-864-3297

(For IL, IN, MI, MN, OH, WI)

NASA Lewis Research Center
Teacher Resource Center
Mail Stop 8-1
Cleveland, OH 44135
(216) 433-2017

(For CO, KS, NE, NM, ND, OK, SD, TX)

NASA Johnson Space Center
Teacher Resource Room
2101 NASA Road One
Code AP-4
Houston, TX 77058
713-483-8696

(For AK, AZ, CA, HI, ID, MT, NV, OR, UT, WA, WY)

NASA Ames Research Center
Teacher Resource Center
Mail Stop TO-25
Moffett Field, CA 94035-1000
415-604-3574